CONTENTS

Appendices

Glossary

The author wants to thank

Dr. Neil Weinstein,
Teresa Holt,
Marion Freund,
Dr. Gerhard Freund, and
Derek Long

for the valuable help.

Section I

What Can You Get Out of This Book?

This book is the opinion of a scientist. The author, in the following called AT, is a physicist and engineer. He has practiced the profession now for fifty years. His PhD is in physics; in addition to this he is also a registered engineer. It is reasonable to assume that his scientific background shaped his opinion on the subject matter of this book. Nevertheless, this book is not an attempt to teach the reader physics since such an endeavor could not be done by one book alone. A library would be more like it. Yet you may safely assume that AT knows his business, and he will only confront you with the final form of the teachings of physics and not elaborate how those teachings where derived.

First and above all, the following question needs to be addressed: Can physics contribute anything to the subject of the book? Normally, one expects that physics is the knowledge of the natural laws concerning matter and energy. However, this statement is not correct. Physics rather is the knowledge of the natural laws concerning energy and *time* (Matter is a property of energy.)

For the subject of the book, it is *time* that is certainly involved with the subject matter. Was there a beginning of time, and will there be an end of time?

The book is not concerned with religion. AT does not claim any expertise in theology. Neither is it an attempt to start a new religion. It is just an opinion. The reader may agree or disagree; in either case, it will stimulate some independent thinking and, therefore, should be worthwhile to be read.

The book is written for the average person to understand. It does not contribute anything new to science. What is presented in the book will not be incorrect, but it may be superficial or oversimplified. The goal is to get the message (the opinion) across with as little ado as possible. A person's opinion is neither right nor wrong; it is just an opinion, with which one can agree or disagree. The goal of the book is not to convince anybody about AT's opinion but to start independent thinking of the subject of the book. And now we reveal a *secret,* namely, what was really the incentive for AT to go through the (considerable) trouble to write this book? The word "secret" is underlined to draw your attention. It is certainly not the money. It is highly probable that money will be lost in this endeavor. Neither is it an attempt to gain scientific glory; as it was clearly stated, the book does not attempt to promote a new theory. It is also not an attempt to convince the reader that AT's opinion is "correct." An opinion is a private matter; agree or disagree with it, but do not argue about it. Have your own opinion. For AT, it was rather disturbing to learn from the few people who have gotten a preview of the manuscript that they pointed out that such and such idea actually comes from Socrates or Plato, etc. Sure, that is even true since during a lifetime, one reads many books and gets influenced by them. The real goal of the book is *to help people face the inevitable.*

Having said all these, AT has learned the hard way that people forget the above underlined statement before they are through with half of the book. *In other words, the goal of the book* is to try to help people form their own opinion and so find out that dying is not such a terrible thing. They rather would like to argue if AT's opinion is "correct." Whether or not AT's opinion is "correct" has no bearing whatsoever on the reader's ability to get some information that helps them form their own opinion. Therefore, AT is forced to take extreme measures to remind the reader on what she or he should get out of the book.

The Roman politician Cato pronounced after each speech he gave, *Ceterum censeo, Carthaginem esse delendam* (in addition, I urge that Carthago should be destroyed). Based on this, AT will have his ceterum censeo printed at the end of each section:

> ***Ceterum censeo, it does not matter what AT's opinion is. What matters is that in this book, you will find explanations of scientific teachings that will help you form your own opinion.***

Section II

How to Read This Book (Definitely Read This, Please)

It is a matter of experience that a person needs to hear a piece of information three times in three different contexts before shehe actually knows it, understands it. (A new English word just has been invented: "shehe" means she or he.)

Therefore, it is recommended that you read this book three times. The first time just cursory, to get some feeling what the book is about. The second time more systematically, and if the interest is still there the third time, the appendices should be read, although these appendices contain some information that goes beyond the subject of this book. What you should not do is to read the last section first. If you do this, you will get the wrong impressions as to what the book is about, and you might decide not to read the book at all.

The book deals with a serious subject; therefore, it is necessary to study it carefully. It also will deal with interesting details, but the book is not aimed to be entertaining. If the reader wants to get an appreciation of the subject matter, shehe will need to work on it.

The world of science is full of jargon. This happened first intentionally (when Greek or Latin words were used) and later unintentionally when English or German words were used having a special meaning only. For example, the English word "power" can mean a lot of things: political power, man power, power of the word, nuclear power, power play, etc. In physics, "power" means *energy* divided by *time.* Not knowing this, the average reader will not be able to make sense out of a sentence that uses this word in its specialized (jargon) form. For this reason, we will indicate all jargon words by *italics*, although such a jargon word will be defined the first time it is used in the text. A more detailed definition can also be found in the glossary.

To help comprehend the subject matter better, there will be after each section a discussion between two characters. One will be AT; the other one will be Mephisto. The name Mephistopheles is one of the many names given to the devil by mankind over the centuries. For the purposes of this book, the devil's advocate will be named Mephisto, and he will attack AT for whatever seems to be a good strategy of objection.

> Mephisto: Do I understand right, you want us to learn physics? You will bombard us with equations that we will not understand and make us look foolish and so you look smart.
>
> AT: Indeed some authors do this in order to impress the reviewer of the book while caring very little if the audience has a hard time understanding the text. But this book is not a book on science but a book describing the opinion of a scientist on the subject matter. AT will make every effort to see to it that anyone trying to understand can succeed. In the few instances when equations are used, they will be translated into plain

> English and their meaning explained also in plain English. What will not be done is to explain how those equations are derived from first principles. Since we deal here with accepted results of established science, you may be assured those equations are correct and need no further confirmation.

Ceterum censeo, it does not matter what AT's opinion is. What matters is that in this book, you will find explanations of scientific teachings that will help you form your own opinion.

Section III

Are Science and Religion Enemies?

It is a basic human desire to gain knowledge as to where we come from and whereto we will be going. In doing so, one believes that this intellectual activity is one feature that distinguishes us from animals.

And indeed if there is a difference, it must be this. It cannot be the use of tools because some animals actually use tools. It cannot be emotions; anyone who has ever owned a dog knows that animals indeed have and display emotions. It cannot be social behavior since many animals display social behavior, and some even have tightly controlled societies. Can the difference be morality? Definitely not. For some species, moral laws are indeed very strict. Some do not commit murder. Granted their moral laws are different from ours, but they do exist.

Summarizing the difference, it is claimed that we are the only life-form that knows there is death. Consequently, we also ask what comes after death. This suggests that humankind is the only life-form that has religion.

Religions should answer the question as to where we come from and whereto we will go after death. Also—closely related to this—what is our purpose on Earth? One may ask whether the answers to these questions given by the various religions are true, even though some are contradicting the others. Yes, they are indeed true—for the faithful, if the faithful are really faithful and truly believe the teachings of their religion. However, the talent of total belief is given only to a few or maybe to none at all. When humankind lost their paradise, they gained some knowledge but lost the ability to believe. They substituted for this the quest for total knowledge to satisfy their need to know what our purpose in life is and what comes thereafter.

It seems to be deeply rooted in the subconscience of humankind that the quest for knowledge is dangerous. This idea can be found in many writings. In the Bible, Adam eats the forbidden fruit. In antiquity, the last veil is not to be removed. In medieval thinking, Faust strikes the Faustian bargain with the devil for gaining unlimited knowledge. Since the expulsion from paradise has already happened, abandonment of the search for knowledge will not bring paradise back. The choice humankind made by starting to search for knowledge initiated an irreversible process.

Yet is science in opposition to religion? Is there a conflict in their respective teachings? Since both claim to teach the truth (the same truth), there should be no conflict. Yet there is conflict. Scientists were burnt on the stake in the Middle Ages because of this conflict. Therefore, the suspicion is that neither science nor religion possesses the total truth. Even if a given scripture were truly the word of God, it may have been transmitted imperfectly. On the other hand, science can indeed explain (naturally) many observed phenomena, but it cannot explain all.

The desire of humans is to learn the true answers to this question as to where shehe comes from and where shehe is going. Religion answers this question by authority; sciences answer these questions by deductive reasoning. Humankind wants these answers, and in case of science, humankind has a right to be given this answer. Even in the case of religion, this right of humankind exists. Scientists have to eat and will eat what others have planted. As a return to society, they will have to provide something these other members of society want. Sure, the scientists can produce gadgets that help them with their chores (but that is engineering). Basic (fundamental) science has no practical application.

Religion offers those answers, and humankind can so be satisfied (saved). Yet this is only true for the true believers and consequently only for few of us. When total belief is missing, compromises have to be made. What science teaches is flawed by the fact that humans perform the scientific research. What religion teaches is flawed as well since the message of religion is interpreted and modified by humans as well. If a final truth is achieved, it should be identical whether it is taught by religion or science.

In this book, we will inquire what science can say about the subject of life after death. One may want to compare this to the teachings of one's religion, and one will see that some features—after some interpretation—are indeed in agreement. If this were not so, one of the two—science or religion—would have to be wrong.

> Mephisto: My boss does not like at all what you are saying, and I submit the pope does not like it either.
>
> AT: I have not talked to your boss lately; as a matter of fact, I do not believe he exists. But if he would exist, what is wrong with what I am saying?
>
> Mephisto: It is the preachers who do not believe what they are preaching. In the Middle Ages, the preachers

committed unbelievable atrocities (the Spanish Inquisition) and got away with it. Only because of the fact that the faithful believed totally were they (the preachers) not decapitated (or worse). The emperor of Germany and the king of Spain (at that time, the same person) did not believe he would go to hell for this. And the preachers believed also there is no hell.

AT: Maybe they knew better. What do you say then? Ask your boss if Emperor Karl (Carlos) is in hell.

Mephisto: This is classified information.

AT: See what I mean!

Then why does religion not teach such a total truth? It actually does, albeit indirectly. Whenever one tries to interpret the teaching of a religion, one should understand that the prime task/problem of any religion is to convince the masses. These masses may be peasants, or they may be computer literates. Obviously, there should be a difference in the wording of the teaching depending on the audience to be saved. Teaching to shepherds the same idea as to computer literates must certainly use different wordings. Shepherds who cannot read, maybe not even count, could make very little sense out of entities like energy, dimensions, and semantic levels. Yet the same idea should be conveyed, namely, that the physical body and reality cannot be the only world there is.

Consequently, the idea of a heaven is widespread and can be found in almost all religions, even the polytheistic ones. It should be noted that the idea of a hell is not so widespread; the word "hell" is not found very often in the Old Testament. Unfortunately, the wording used, when teaching the ideas a certain religion is based on, needs to be changed. This did not occur to the practitioners of religious authority. Or more likely, they knew very well that the advances of science would require modifications in their wording. Rather than changing, they decided to consider science as the enemy and try to suppress the findings of science. The most striking example is given by the incident when Galileo's clerical friends refused to look through his newly invented telescope because they were afraid to see something that is against their teachings.

That science is the enemy is also made clear in the saga of Faust. This is folklore, but somebody must have invented it back around the year AD 1,000 who understood what was going on. Faust sells his soul to the devil for unlimited knowledge. Of course, this saga is an underground social and political protest. The object of the protest is the claim that the ruling power structure, the king or the emperor, is sent by God and rules in behalf of God. To make his idea stick, the people had to be kept in the dark, meaning deprived of the results of science. Or even better, scientists had to be kept from getting any inconvenient results. Those who did, they would burn since they obviously made a pact with the devil. In short, any modification in the teachings of religion enforced by increased scientific knowledge would pose a severe threat to the idea of the king being sent by God.

Basically, one has to conclude the practitioners of the religious authority did not believe themselves what they were teaching, namely, that the king is sent by God. Actually, they knew better. Then why did they teach it anyway? They knew that they did not contribute anything to the economy in form of work; they needed the worldly power structure to feed them and protect them.

The argument that there is a higher power who created the world, meaning now the physical world—reality—is indeed a convincing argument, although there is, of course, no proof. The atheist may say the world (reality) was always there. The big bang was only a rearrangement of the chaos. And indeed even the Old Testament states, "In the beginning was chaos (*tohu wa bohu*)." God created the world, but the chaos was there first.

Humankind may be inclined to go along with the idea that the soil shehe steps on is just there and was always there. An animal certainly would have such a notion. However, there is another notion that was being questioned at the instant humankind changed from animal to human. This is the notion that there is more than the soil one steps on; there must be another world whereto one goes after death. Once this notion was accepted, these animals had become humans and started to bury their dead. They also left messages behind of the deeds of the deceased in form of drawings on the walls of the caves.

While the inanimate world can be taken for granted (maybe as a result of residue left behind in our brain), humans find it very hard to accept that an animalistic world of ideas, meaning intelligence, formed by itself. This is true for all humankind, and therefore, there are really no atheists. Those whom the religious authorities consider atheists are persons who do not believe in the personification of the power that created the world as this particular religion teaches and as specified by their particular detail.

The major disagreement of the so-defined atheist and the religious authority is the question of continuous divine intervention into our personal lives. If there is no such intervention, praying is useless, as well as any organized assembly of people who does the praying.

Ceterum censeo, it does not matter what AT's opinion is. What matters is that in this book, you will find explanations of scientific teachings that will help you form your own opinion.

Section IV

A Model of the World

A. The Higher Semantic Levels

If we are looking for another world to go to after death, we need to understand the structure of the world we are living in as best as we can. Therefore, in this section, we will develop a model of our world.

What is a model? Why do we need a model of the world? Why can we not describe the world as it is? To answer this, we might use the popular metaphor of the three blind men who examine an elephant. One of the three, feeling along the elephant's tail, concludes that it is a snakelike thing. However, the second one claims it is something like a tree since he touched one of the legs. Yet the third one finally pronounces it must be a wall because he examined the enormous body. On comparing notes, they cannot reconcile their research results. Something cannot be a wall, a snake, and a tree at the same time.

Modern physics experiences a similar predicament. Some experiments "prove" that light is an assembly of particles while other experiments "prove" that light is a wave. "Common sense" tells us it has to be either one or the other. It cannot be both. To see how physics solves this problem, consult appendix 5 for more details.

For the time being, let us return to the three blind men. How would they solve their problem if they were blind yet also wise? They might conclude that the object they are trying to describe is too complex for their senses, meaning that they are incapable of describing it in its totality. This is indeed true because even if somebody told them the elephant is gray, this would not mean anything to them since they are blind, they cannot possibly envision what "gray" is. Consequently, they might conclude they have to speculate on information they do not have. Unfortunately, they could not comprehend the results of their speculation even if they speculated correctly. Speculating on information one cannot comprehend does not really help. What one has to do is to modify (simplify) the unknown, complex information until it is in a shape that one can comprehend. This is accomplished by formulating a "model." In this case, the model would be a simplification of the complex entity "elephant."

What is a model? As it is well-known, a model of a ship is not really a ship. It exhibits only some of the properties of the real thing "ship." The model looks like a ship, but it cannot transport people. It is important to notice that a model does not necessarily have to be smaller than the real thing. A model of a molecule has to be larger than the real thing; otherwise, one could not see it. The word "model" implies that some properties of the real object are represented, but not all. This allows omitting properties that are too complex for the person using the model to comprehend.

One may say that in the above example of the blind men, this is fine, but regular people can see and so comprehend that the elephant is an elephant. Not quite. Even regular people have limited senses. They cannot "see" whether or not the elephant has a soul. If he behaves strangely, they cannot see if he is mentally disturbed or just plain angry or excited. Therefore, seeing people also have to construct models concerning properties they cannot sense. In our case—the model of the world—these additional properties seem to be more complex than those the three blind men encountered. This is so because humankind conceived this metaphor so that "normal" people can understand their limitations.

In order to use the proper jargon, we introduce now the concept of *higher semantic levels.* The word "semantic" suggests that this has to do with language. If one were convinced that for logical thinking a language is necessary, obviously the more complex the language is, the better the logic. It may be debatable whether or not it is true that a language is needed for logical thinking. The present book is not the proper platform to debate this. Here, it is assumed to be true since it helps explain a difficult subject. It should be also recognized that the language of physics is mathematics. Indeed, mathematics is defined as applied logic. Mathematics helps the logical thinking process enormously, yet it can really compare only one existing logical conclusion with another one. It cannot make conclusions by itself. (This is only true for conventional mathematics—e.g., calculus—where everything evolves around an equal sign. Other mathematical systems are beyond the scope of this book and would not be germane to it anyway.)

Nevertheless, if we encounter observations that cannot be handled by our model of the world, we state that *higher semantic levels* are needed to overcome our problem.

B. Theories and Facts

Physicists make observations of nature and then need to explain what they have observed. The physicists of the antiquity observed earthquakes. So they came up with a theory, namely, that Earth is flat and rests on the back of an enormous elephant. The theory was supported by the observations of the three blind men. All of a sudden, their observations made sense. Now they can explain why earthquakes happen. Each time the elephant lifts one leg, shakes it, and puts it back down, Earth is disturbed and shakes. Great. This is the explanation for earthquakes. Now they need further confirmation for their theory. Why does the elephant lift his leg? He shakes it to scare the mosquitoes away. Therefore, in a year when there are more mosquitoes than normal, there must be more earthquakes than in a year having a smaller number of mosquitoes. Nature sometimes comes up with weird coincidences; maybe the mosquito count supported the theory of the ancient physicist.

From this example, we can see the difference between theories and facts. Facts are facts, and theories are proposals as to how the facts need to be in order to explain the observations. Theories are never wrong or right; they are only consistent or not consistent with observations. This means they are only useful or not so useful. There will be always a time when a theory is no longer useful because a new observation is made that contradicts this theory. E.g., there was a year when there were no mosquitoes around, but there was an earthquake. Too bad, but we can fix this: the elephant has one leg that sometimes falls asleep. Then why do earthquakes happen all over the world and not only on that part of the world that is located on top of the leg? And so on and so on.

But how can we recognize a fact? The truth is we cannot. For example, we realized we live in a three-dimensional world.

But is there a fourth dimension? Using mathematics, we can compute how many corners, edges, and areas a four-dimensional body should have (see also appendix 2 on dimensions). But where is the fourth dimension? To tell us this constitutes the same problem as explaining to a blind person what the color green looks like.

Now our grand conclusion: we will never know the total truth as to what the real facts are *because of our limited semantic levels*.

C. The Different Regions of the World

Our model of the world states that, like any complex structure, the world can be divided into different substructures. For the sake of easier discussions, we name the different regions of our world as follows:

1. Region of Reality
2. Region of Existence
3. Region of Infinity

The *Region of Reality* is the world we live in, something that we can see and feel. The *Region of Existence* is a world where ideas, laws of nature, information esthetics, etc., reside. We cannot see or feel this world, but we know it exists. Finally, there is the *Region of Infinity*, which we cannot imagine at all. Not only can we not feel or see this part of the world (*infinity*), we cannot even comprehend it. We may be able to describe it in the form of metaphors, but these analogies would be imperfect because we are trying to describe something using a language that does not have the necessary vocabulary (semantic level) for an adequate description.

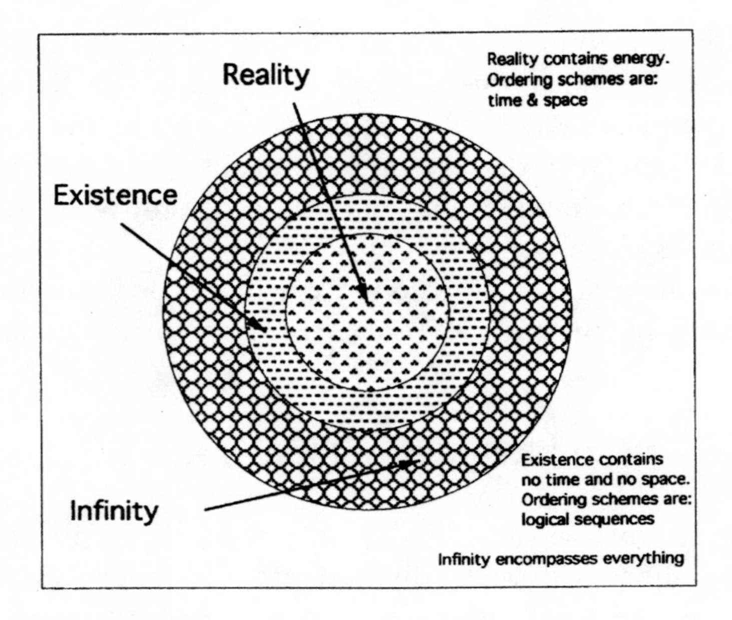

Figure 1. The Regions of the World

We define the *Region of Reality* as the part of the world that can be detected by our senses or by measuring instruments and scientific sensors. The content of *reality* is *energy*, and this content is organized by the ordering schemes *space* and *time* (see also appendix 1 on energy and appendix 2 on dimensions).

We see something because it emits light, which is a form of *energy*, and our eyes absorb this light with the retina. In principle, the same is true for all our other senses or for scientific sensors. In this case, different forms of *energy* are used in the exchange process. We discuss *energy* in more detail in appendix 1. The sensing process performed, by either our senses or by scientific measuring devices, is accomplished by an exchange of *energy* between the object sensed and the sensor.

What is discussed in the two appendices referenced above shall be summarized here briefly. It is claimed that *energy* is a *substance*, and it is the only substance there is. All *energy* is endowed with mass. *Energy* appears in different forms; one form of *energy* can be converted into another form, but it is impossible to create or destroy *energy*. This fact is known as the first law *of thermodynamics* or as *conservation of energy*. For example, the process that "generates" or "produces" electricity in the local power plant is actually a conversion from heat (steam) to electricity. AT also claims that there is a fifth law of thermodynamics that states that at each energy conversion, all three forms of *energy* take part. In other words, at each conversion event, heat is generated.

Summary: The *Region of Reality* is filled with *energy* and is structured by space and time.

> Mephisto: Let me interrupt right here. You promised to use only accepted results of the official science for this book, and now you invent a brand-new law that is pure speculation.

> AT: No, it only looks like this. The practitioners of physics and engineering always knew that there is friction in most of their applications. The only new thing is to give that phenomenon a name.

> Mephisto: But it is wrong. For example, the Moon, when circulating the Earth, continually converts Potential Energy into Kinetic Energy and vice versa because the Moon's orbit is not circular but elliptical.

> AT: You got this right that the Moon has an elliptical orbit and constantly converts Potential Energy into Kinetic Energy and back to Potential Energy again. However, this causes the tides and, therefore, waves in the ocean, which eventually die out and their energy

> becomes heat. However, the major cause of the tides is rotation of Earth itself.

We define *existence* as the part of the world that is filled with ideas, knowledge of the laws of nature, logical relationships, etc. To bring these various idealistic entities under a common heading, we call all these *information.* Therefore, *reality* is filled with *energy* and *existence* if filled with *information.* Anything that can be described is part of *existence.*

Information can be expressed or described by words, images, numbers, mathematical equations, or other media we have yet to conceive. An important feature of *information* is that it is indestructible. For example, all the books that contain a certain truth, such as the Pythagorean theorem, can be destroyed, yet the theorem continues to *exist*. It is still true that the sum of the squares of the two shorter sides of a right triangle is equal to the square of the longer side, even though the books that describe this have been destroyed. Or a character in a book like Scarlett O'Hara in *Gone with the Wind exists* although she never lived as a *real* person. These examples indicate the limits of the *Region of Existence.* Anything *existing* has to be *describable*. If it is not describable, it is non-informatio*n* and, therefore, does not *exist* in the *Region of Existence.* The rules for "describability" are that the description cannot lead to logical contradiction. In short, *information is the knowledge of everything that ever happened or could have happened and everything that may happen.*

One might ask, where is this *Region of Existence* located? That is certainly a valid question. However, the word "located" is related or reserved to three-dimensional space, which is the *Region of Reality.* An object or a person is "located" in space and time.

The *Region of Existence* is a region of the world that is different from *reality*, and it is, therefore, certainly not located in *reality*. Then where is it? (Not using the word "located" anymore.) The answer must be: it just *exists*. Or it is everywhere. The *Region of Existence* is not endowed with energy; therefore, it cannot be seen or detected with scientific instruments. As an example, one may say a thought cannot be seen or be detected with instruments. One can detect that a person thinks, but one cannot detect the contents of the thought, e.g., the great idea or a new melody. However, we certainly must admit that thoughts *exist.* We might claim that thoughts rest in the brain; therefore, we know where they are located. However, that is not true. The thinking process may be located in the brain, but the product of the process (for example, the new melody) is certainly not, or at least not exclusively. The new melody will exist forever even after the composer has died. More precisely, this "new" melody already *existed* since the beginning of time (if there is such) and will *exist* till the end of time (if there is such). What the composer did was to *realize* it, to make it available to humankind. In our terminology, we say it was transferred from the *Region of Existence* to the *Region of Reality*—was *realized*—through the brain as a receiving station.

Is the *Region of Existence* organized in some form of order analogous to the *Region of Reality*? Yes, it is, although the order is not *space* and *time*. It is important to remember that *time* is not part of *existence*. The contents of *existence* are information bytes that are coupled to one another by their logical contents. Later on, we will see that the length of such a logical train is important for life after death.

One might ask if God resides in the *Region of Existence.* To prove the existence of God was an intensive effort during the Middle Ages. Modern theology has concluded that such a proof is impossible. This actually makes sense, if there were such proof; his existence would be a fact of knowledge. Once one knows something for certain, there is no longer a need to believe. However, belief is the basis of any religion. Upon closer examination of our definition of the *Region of Existence*, one can see that this definition does not state that God *exists*. The reason can be found in the condition that anything *existing* has to be *describable.* However, at least for the three large monotheistic religions, God is almighty. This implies that he is also limitless and infinite (therefore, indescribable). Therefore, if we believe that God is part of our world, we must conclude that he is everywhere in *reality*, in *existence*, and in *infinity*.

This book is not about religion. We make this comment here only to point out that our model of the world, which is based on philosophical/physical theory, indeed has room for religion. This is in contrast to the common belief that the physical sciences are atheistic.

We define the *Region of Infinity* as that part of the world that we cannot comprehend (due to our limited semantic levels). We like to believe that the universe extends into infinity. This means the universe is infinitely large; in every direction we choose to go, we would find no end. If the universe had a boundary somewhere, we would ask, is this a wall? If so, then what is behind this wall? On the other hand, we cannot comprehend how something can be infinitely large. As human beings, we are taught that nothing is of unlimited supply. We cannot comprehend that the world might be limited, but we neither can comprehend that the world might be unlimited. This incapability of comprehension includes the fact that infinity is not describable. Anything that is describable must have properties since these are necessary to describe entities. If something has certain properties, it must be lacking others; otherwise, a distinction cannot be made. Therefore, if God were part of the *Region of Existence*, he would have to have certain properties and, therefore, would lack other properties and, therefore, would not be almighty. In other words, God is incomprehensible.

Mephisto: I cannot find any physics book that contains a figure that resembles your figure of the regions of the world. Did you again invent something new and try to sneak it in?

AT: Not exactly. Physics is the discipline that is concerned with energy and time. We call it here the *Region of Reality*. The author of a physics book does not need to declare on the first page that physics is concerned with energy and time. He can expect that the reader knows this. Neither does he need to mention that there are other entities that are not the concern of physics. The fact that items "exist" and the fact that there is "infinity" is not a concern of physics, but on the other hand, nobody can reasonably claim that there is no infinity and that there are entities that have no mass, no energy, and no time.

Mephisto: Fine. Just tell me where is the "*Region of Existence*."

AT: I hate to admit this, but that is a good question. Therefore, I will try to answer it. Imagine you dream something. It could be persons you never have met or landscapes or interior of rooms. What you have dreamed is describable; therefore, it belongs to the *Region of Existence*. Now the night is over, and the dream has disappeared; you might have forgotten most

> of it. Let us only talk about this part of the dream. Where are these people now? They are certainly not in your head, but they exist. They are certainly not at any corner of the globe or on the Moon or Mars. They cannot be detected since they have no mass; neither are they linked to time. The answer to the question, therefore, is they are everywhere but not in the space-time continuum that we call the *Region of Reality*. You may blame our limited semantic levels for this. If you were Mephistopheles instead of Mephisto, you could have asked with the same validity, where is hell?

Now, a few more words about *infinity*. Infinity is infinitely large. To amplify this somewhat, we write the following:

1. $x = n * x$
2. $\infty = n * \infty$
3. $0 = n * 0$

The *n* means any number less than infinity. The "*" is the sign for multiplication (instead of a period or an *x*). Of course, the same amount should stand on each side of the equal sign. In equation 1, this is only true if $n = 1$. Therefore, this equation can be considered as the definition of the function of the equal sign. Equation 2 is true for *n* being any number less than infinity. In the extreme, *n* could be equal to infinity. But according to mathematical convention, such an equation is not considered legal. What we are trying to say here is that infinity is so large that it cannot be made larger by multiplying it with anything. This sounds very weird. (Some mathematical systems do not agree with us, neither do we agree with them.) To make a weird statement more understandable, we write the third equation. This equation states that one can multiply 0 with anything, and it will not get bigger than 0. Well, this is easy enough to understand; nobody will claim that this is weird.

D. The Points That Section IV Was Trying to Make

The major point that this section is trying to make is certainly that the world is more complicated than just the three-dimensional *reality* we live in. The fine (minor) points are distributed all over the section. This was necessary to put these fine points in the proper context for better understanding. One of the fine points of section IV is actually the most important point one can get out of this section, namely, that the *Region of Existence* is *not* structured by *time* and *space*. The consequences of this are enormous albeit it took a lot of words to bring this out. Plainly stated, there are parts of our world that are eternal and having contents that are eternal. If there would be a way to make a claim that the world of existence contains souls, an argument could be made that souls are eternal. After the stars in the universe have turned into novae, neutron stars, black holes, or whatever; the contents of the *Region of Existence* are still there. For the rest of the book, maybe a reasonable argument can be established for a possible claim that souls "*exist*."

> Mephisto: Whatever you say about souls, my boss got them. Better be careful, you may be one of them.
>
> AT: Your contributions to the book are most of the time not very helpful, but this time they are. You admitted that there are souls, albeit only bad ones. But where is hell?
>
> Mephisto: That is classified information.

AT: That may be so, but considering this a little bit, we find out that we are already in hell. Could there be anything worse than what humankind did to itself since recorded history?

Ceterum censeo, it does not matter what AT's opinion is. What matters is that in this book, you will find explanations of scientific teachings that will help you form your own opinion.

Section V

The Structure of the Three Regions

A. The Region of Reality

As pointed out above, each of the three regions has a certain structure. This means they display properties that make them distinguishable from the other regions and from chaos. As we will see, *reality* is highly structured; *existence* is less structured, and *infinity* has no structure at all.

We discuss the *Region of Reality* first. It was said before that *reality* is endowed with *energy* that resides in *space* and *time.* It is commonly assumed that *space* has three dimensions (see appendix 2 on dimensions). It has also been speculated that space is bent into a fourth dimension, but we cannot imagine where this fourth dimension might be (because of our limited semantic levels). This is also covered in some detail in appendix 2.

The *theory* of *relativity* teaches that *space* and *time* form an inseparable unit, usually called the *space-time continuum.* To make the present section more easily understandable, let us begin by summarizing what is explained in more detail in appendix 1.

It is stated there that *energy* is a *substance*, and it is the only *substance* there is. *Energy* itself can assume different forms. The three main forms of *energy* are

Potential Energy,
Kinetic Energy, and
Random Energy (Heat).

Potential Energy extends into space. *Kinetic Energy* extends into time, and *Heat* extends into both. All forms of *energy* are endowed with mass. The most misunderstood equation in physics is $E = mc^2$. This equation is called the mass-energy equivalence, not the mass-energy equality. This equation should be properly written with an identity sign (which is three lines on top of each other). The interpretation of this equation is that each form of energy is endowed with mass (see more on this subject in appendix 1). In this appendix, it is also explained that *energy conversions* need to occur for things to happen. *Energy conversions* mean the conversions from one form of energy into another one. (When you drive a car, you convert the chemical energy in the fuel [Potential Energy] into velocity [Kinetic Energy].)

According to our definition of the *Region of Reality*, space and time are *real*, but only if they are filled with *energy*. *Therefore*, any space that does not contain energy *exists* but is not *real*. All observable space must be filled with *energy*. The very fact that it is observable implies that it is at least filled with light (a form of *energy*). Any observation consists of converting one form of energy into another one.

In figure 1 of section IV, the three regions are shown. According to this figure, the *Region of Reality* borders the *Region of Existence*. This implies that if there should be no more energy (galaxies) in the very far distances of the universe beyond the last observable galaxy—billions of light years away—the *Region of Reality* crosses over into the *Region of Existence.* There is no wall.

However, some theories say that the three-dimensional space is bent into a fourth dimension. Therefore, a spaceship traveling in a straight line would eventually end up at the place from which it started without ever leaving the *Region of Reality.* Remember, theories are never wrong or right; they are useful or not so useful. What the facts are in this case, we probably will never know. What you are reading here is an opinion, and this opinion does not deal with a fourth dimension. Later on, you will see that a border between the two regions is very important for other metaphysical explanations.

The structure of the *Region of Reality* needs to be discussed here only shortly. The *theory of relativity* teaches that there is no absolute space. This means that one cannot designate a certain point in space as the center of the universe. However, one can designate a certain point in space relative to a mass in space (e.g., a star). Empty space cannot be measured; one can only measure the distance between two masses (objects). Then what is the three-dimensional space anyway? If empty, it just *exists*; if it contains objects, it is ordered by ordering schemes (*space* and *time*). These ordering schemes enforce the law of causality.

This can be shown by the following example. Take two identical billiard balls and shoot one onto a second one that is at rest. If the first one hits the second one head-on, the first one will stop, and the second one will take off (provided they have the same masses). What happened? The first one had Kinetic Energy (it moved). The second one was at rest; therefore, it had no Kinetic Energy. After the collision, the second one moved, and the first one is now at rest. An *energy conversion* has taken place. One could ask if it were possible that the second ball could take off before the first ball hits it. If the first ball would come to rest from where the second ball took off, then the *law of conservation of energy* would *not* have been violated. However, the law of causality would have been violated. No effect can take place unless there is a cause for it. The ordering scheme "time" sees to it that the cause comes before the effect.

It is also required that the balls touch each other; otherwise, the second ball would have moved without a cause. Touching means a part of each ball occupy adjacent points in *space* for a short time. The ordering scheme "space" enforces this. Consequently, the structure (*time* and *space*) of the space-time continuum enforces the law of causality.

It is interesting to observe that *time* cannot be measured and neither can the empty three-dimensional space be measured. How can we say that *time* cannot be measured? There are clocks. To make it as simple as possible, we "measure" *time* with a pendulum. When the pendulum is deflected to one side, it sits still (for a very small moment); therefore, it has then only Potential Energy. When it is on the bottom of its swing, it has only Kinetic Energy (it moves). When it is on the other side, it sits still again and has only Potential Energy. Therefore, *energy conversions* have taken place. In order to "measure" how much time has taken place between two events of interest, we count the number of *energy conversions* that have taken place or, in this case, how many pendulum swings have taken place. All we can do is to assume that each of these *energy conversions* take the same time to happen.

According to the above argument, *space* and *time* are ordering schemes of the *Region of Reality* but not of the *Region of Existence*. This means the contents of *Region of Existence* are *not* space and *not* time.

B. The Region of Existence

Now let us discuss the structure of the *Region of Existence*. Our definition states: "What can be described exists, but if it is endowed with energy, then it is real." There is no limit to the number of things that can be described. Existence is unlimited while reality is limited since there is a limited, albeit large, amount of the substance *energy*. Yet it is important to notice that *existence*, while unlimited, is not infinite because there are things that cannot be described, because the description would lead to logical contradictions. For example, one cannot place a rectangular table in a corner of a round room. Such a situation does not exist. In other words, existence has (logically) consistent properties. What fills the *Region of Existence* then? We have said that the *Region of Reality* is filled with energy. As an analogue to this, we claim that *existence* is filled with *information*. What energy is to reality, information is to existence. There may be an unlimited supply of information but not an infinite supply because there must be some information, which is not possible (because it would lead to a logical contradiction). For this reason, there must be voids in the structure of existence. The structure of the *Region of Existence* consists of all possible—that is, describable—information. In comparison to existence, the structure of reality is more rigorous. Reality contains space and time. We claim that space and time are ordering schemes and, therefore, provide structure. Since there is no *time* and *space* contained in existence, any order in the structure of existence needs to consist of logical consistency.

As we claim that *energy* is a *substance*, we claim *information* is an *entity*. *Information* resides in *existence* while energy resides in *reality*. While space and time is familiar to us by physical experience, *existence*—the host of the entity "information" is not familiar to us. We cannot point out a location for it because it has none. *Existence* just *exists*; it does not need a location to reside in. There is the process of *realization.* Only after *realization* has taken place, images (copies) of *information* now reside in space and matter. The ordering schemes for reality are space and time. The ordering scheme for existence is *logical consistency.*

For something to *exist*, it has to be describable. Something that is indescribable does not *exist*. What do we mean by describable? A description could be a mathematical one, e.g., a mathematical equation. It could also be a set of logical sequences or a set of properties.

Therefore, a circle could be described by an equation:

$$r^2 = x^2 + y^2.$$

Or it could be described as a set of logical sequences: "A circle is the location of all points that have a common distance from a given point, namely, the center." It could also be described in terms of properties: "A circle would be void of corners, would be continuous, and would possess the same curvature at all points." According to either of these descriptions, a four-cornered circle does not *exist*.

Reality and *existence* have an important connection, namely, the process of *realization.* In nature, there seems to be an *urge for realization.* What is this? We are using the word "urge" for a phenomenon, which we infer to be present but cannot be detected or be described. We have an inkling, but not a description. To attack the problem, let us talk about *realization* first. What we described above concerning the circle should illustrate this fairly well. The physical circle, drawn on a piece of paper, is the realization of the idea "circle." *Realization* must mean to image (copy) an *entity* from *existence* into *reality.* How is that done? Can it happen by itself?

Now comes a claim that is the most important message of this book: *Only life can perform realization of information.*

There must be somebody who draws the circle. There must be somebody, who appreciates the information expressed by this circle. Of course, one could say this is not so. For example, take the three-dimensional analog of the circle, the sphere. Nature grinds unshapely stones into spheres by the action of water in rivers. Is there somebody who appreciates the spheres? No, of course not. But the spheres roll easier. They may be used to carve a better riverbed to help the river find the ocean so that the area is drained and plant life can sprout. One may call this serendipity or more modernistic synergy. Yet such a seemingly unplanned cooperation really is a precursor of life. Therefore, at this point, we propose here that the *urge for realization* must have been the reason why life was created.

Naturally, there are many mechanisms for *realization.* Drawing is one of them, but there are many others. Whatever life does is dedicated to *realization* of *information*. In order to express a piece of information, energy must be rearranged. So to speak, the law is hewn into stone. The "law" is *information*; the "stone" and the process of changing it is *energy*. If we look at the development of civilization, we see how humankind made progress by first learning to harness energy and then by learning to manage information. We may think humankind does this for its own good. However, we claim here that it is the *purpose* of life to do this.

The next obvious question is what is *realization* of *information*? The easiest way to understand this is to use the written word as an example. A set of written words expresses an idea (is information). The same idea can be written in different types of letters or even in different languages. In whatever way the realization is done—hewn in stone, written on paper, or appearing on a word processor monitor—in either case, a material medium is needed to do this. It is either stone and the energy to change it, or it is paper with the ink on it and the energy required to print it, or it is electrons hitting the screen of the monitor and the energy required to manipulate them. In other words, *realization* means to bring an idea (*information*) from *existence* into *reality* by changing a part of reality to reflect this particular idea.

The fact that a few grooves in a stone can mean something is truly amazing. What we are saying here is that mass can have shape, and this shape contains information. Since *mass* and *energy* are equivalent, *energy* must as well have shape. This became clear to us when we learned that radio waves could be modulated to carry information. We like to consider this fact—that mass and energy can carry information—as a demonstration for the necessity of life. It takes the presence of an intelligence to accept the meaning of these shapes and to act on it. The intelligence (and its supporting system, the body) was created to facilitate *realization.*

After *realization* has occurred in our brain, something in our brain (the nerve cells) has been modified. The nerve cells are made of matter, meaning they belong to the *Region of Reality* and are not the information that rests in the *Region of Existence.* What is in the brain is an *image* of this information. Similarly, if such information is written in a book, the arrangement of the letters that represents this information is only an image but not the information itself. The information in the *Region of Existence* is not related to space and time; *therefore, it is eternal and indestructible*. The image of information resting in our brain is destroyed when we die.

Basic particles (atoms, electrons, neutron, etc.) carry no information since they have only a limited amount of properties (namely, mass, charge, and spin) but not a well-defined diameter or a certain color or even shape. Molecules can carry some information. Large organic molecules can go up in complexity as far up as to the genes, which carry the genetic code.

Of course, of utmost interest to us is the storehouse for information, namely, our brain. Once we understand why our brain (and, therefore, *we*) can comprehend information, *we will have some understanding how this information gets there and whether or not it goes somewhere after the brain ceases to function*. This will help us form an opinion on life after death.

C. The Region of Infinity

Next we discuss the structure of *infinity*. Based on the above statements, it seems only logical to claim that *infinity* has no structure at all. Or, more carefully worded, it has no structure that human beings can comprehend. For human beings, *infinity* is all encompassing, unlimited, and incomprehensible. Sure enough, mathematicians have devised different structures of infinity, but the equations describing such structures have no meaning that can be comprehended by human beings.

Where is infinity located? Above, we pointed out already that "location" has only a meaning when applied to the 3-D space. Since *infinity* is incomprehensible, there is certainly no way to comprehend a place where it may be. Yet we know that there must be infinity.

Following this enthusiastic—but noncommittal—statement, the time has come when one needs to ask, Are we sure that there is a *Region of "Existence"* and a *Region of "Infinity"*? The answer is yes, we are sure, but we have no proof. Maybe we have something close to a proof, and this is the reason why we answered the first question with yes.

Euclid, an ancient Greek, coined the word "axiom." An axiom is, for example, the statement that two parallel lines never cross each other. This cannot be proved since they could cross in *infinity* to which we have no access. Yet there is really no doubt that in *reality*, these two lines never cross. Such a situation is meant when it is claimed that a certain fact is an *axiom*.

We claim here that *reality*, *existence*, and *infinity* are axioms in our description of the world.

Ceterum censeo, it does not matter what AT's opinion is. What matters is that in this book, you will find explanations of scientific teachings that will help you form your own opinion.

Section VI

The Law of Causality

The *law of causality* is the basis for all science. It states that any effect that happens, happens for a reason, meaning it has a cause.

The laws of nature dictate the relationship between a "cause" and the resulting "effect." These laws of nature apply at all times without exceptions. This enables science, after discovering the laws of nature, to predict the outcome of future effects. For instance, the law of gravitation will "cause" the apple to fall to the ground (an "effect"). Our conviction is that this will happen each and every time an apple is released. We do not believe that in one out of a hundred million effects, there will be an exception, and the apple will not fall or may in fact even rise.

Only if such natural laws exist, which have no exceptions, is serious science possible. If something should happen without cause or if something should happen in violation to the outcome expected based on the laws of nature, we would classify this as a miracle. This conviction is so strong that if only one exception from a law of nature were observed, we would conclude we formulated this law of nature wrong rather than admit that there are exceptions to any law of nature.

It should be pointed out here that modern physics abandoned (so to speak, rescinded) the law of causality (albeit only for the interaction between microscopic particles). We need to mention this already here although we need to discuss the consequences of this abandonment in an appendix in more detail (see appendix 5 on the probabilistic character of the laws of nature). What is said in this appendix can be summarized as follows:

Modern physics claims that the laws of nature do not predict what actually will happen with absolute certainty. These laws rather give a probability for a certain "effect" to happen. It is taught that violations of the law of causality happen practically only for the behavior of microscopic particles and become less and less pronounced the larger the mass of the particle concerned is. Therefore, for macroscopic particles, the laws of nature apply (almost) exactly.

Does this mean, when applied to the example of the apple, that occasionally the apple may rise instead of falling (albeit very seldom)? Experiences tell us that this is, of course, nonsense. The problem is that there is no sharply defined limitation as to whether the laws of nature apply exactly and when not. That may sound nit-picking, but the consequences of such teaching are enormous. It has to do with the existence or nonexistence of a free will. Without a free will, any religion would become meaningless. This topic is discussed in more detail in appendix 5.

Why is the (alleged) breakdown of the law of causality so important for the subject of this book? The answer is that the law of causality states that the outcome of every event that will happen is determined unambiguously.

Events fill our lives. A heartbeat is an event, a birth of a child is an event—everything that happens is a series of events. Correspondingly for each event, there is a cause—a reason—

why this event happened, and each event that has happened is the cause for another event to happen. These resulting events are, therefore, not only predictable; they are also unavoidable. This simple sentence says that the future is predetermined and unalterable. In other words, if millions of years ago even only one pebble on the beach would have been at a spot different from where it actually was, the whole history of the world would have turned out differently. This is so because each event causes the next event to happen. Therefore, the chain of events would have been different from what actually transpired. Consequently, one is tempted to state: Everything depends on each other, and even a small change in the chain of events will lead to a different outcome down the line.

The school of philosophy that subscribes to the above reasoning is known as determinism. Of course, this is incompatible with most religions because it negates free will or the capability of humans to make moral decisions. The probabilistic character of the laws of nature allegedly restores this free will of an individual, even though the probabilistic feature applies only to microscopic particles. One might claim that the process of thinking is linked to changes in small electric charges in the brain cells, which are in the last analysis determined by the movement of electrons. Admittedly, to reconcile the uncertain (indeed unpredictable) behavior of electrons in the brain of an individual with an action of free will of this individual is, to say the least, problematic. Nevertheless, one can at least see that the introduction of the probabilistic nature of the laws of nature nullifies the necessity to believe that everything is predetermined. Now one can conclude that some events happen at random and without proper cause, and therefore, the future is not predetermined.

However, let us interject here that we will show in a later section that there are ways for explaining the existence of a free will in other ways than repealing the law of causality.

The probabilistic nature of the law of causality also poses a dilemma for our model of the world. On the one hand, major events are predictable and predetermined. The solstice scheduled to happen two hundred years from now will indeed happen and is unavoidable. On the other hand, a mental decision of a world leader is unpredictable if it is subject to the free will of that particular individual. Once his decision is made one way or the other, it may have profound effects on the outcome of many other events including the fate of other people.

For the purposes of this book, we like to call the fact that a minute event, like a small change in an electrical charge of a brain cell, causes an effect of huge proportions, the *trigger effect*. The existence of a *trigger effect* suggests that the future is not totally predetermined and unalterable but can be influenced by moral (or not-so-moral) decisions of human beings. Basically, this applies to all thinking beings. A lion may be in a nasty mood and kill a man on a safari who, if allowed to live, might have become a dictator and might have caused the death of millions of people.

> Mephisto: I am relieved that people have a free will. Without it, my boss would have no business. But if it is claimed that the free will is caused by the unpredictability of the movement of some electrons in the brain is not really the exercise of a free will of the person. If anything has a free will in such a situation, it is the electron.

> AT: It is a fact that can be experimentally proved that minor deviations in the path of an electron are only predictable within a certain probability. That is all I am saying. The exact connection to a free will may here be treated too superficially. See in a later section

in the book how the *feeling* of *I* is connected to and influenced by the *Region of Existence*.

Ceterum censeo, it does not matter what AT's opinion is. What matters is that in this book, you will find explanations of scientific teachings that will help you form your own opinion.

Section VII

The Eternal Soul

A. The Feeling of I

Many religions, contemporary ones as well as ancient ones, teach the belief that there is an eternal soul. To form an opinion whether there is indeed such an entity is very difficult. However, if there is one, it needs to be part of the *Region of Existence* since it has no mass or shape. If it had these properties, it could be detected by scientific sensors, which obviously did not happen so far. As an entity resting in the *Region of Existence,* it is *eternal* and i*ndestructible*. Furthermore, it has no relationship to location (*space*); therefore, it cannot be claimed to rest *inside* the mortal body.

The above statement is extremely important concerning the subject of this book. Therefore, some more discussion will be in order. For this purpose, we invent an entity that we call the *feeling of I* ("I" like ego), which is very close to what a soul might be if there is a soul. The *feeling of I* is the feeling or knowledge that we are here. The *feeling of I* does not easily go away. If a person has amnesia, shehe still has that feeling, despite the fact that a major part of the memory is no longer there. (It still is there; it just cannot be accessed anymore.) Does a person in a coma still have a *feeling of I*? Yes, this person does if shehe comes back from the coma. However, this person does not remember what happened during the coma.

Some of us remember the time when we suddenly recognized that we are here. This happens at an age of two or three. Obviously, we had a *feeling of I* before, but at this instance, we remember that we are here already for some time. Prior to this time, we lived and did things to stay alive (e.g., eating). If we had a *feeling of I*, it must have been a very primitive one.

If such a stepped-up organization of the *feeling of I* exists, then does a dog have such a reduced version of a *feeling of I*? One is tempted to agree. However, this would lead to the conclusion that a dog also has a soul (albeit a primitive one). Then one would have to go on and conclude that all animals have souls. This is slippery ground and is not necessarily germane to the subject of the book. Nevertheless, the *feeling of I* needs at least a part of the brain to function. Consequently after death, the *feeling of I* is destroyed.

Summarizing what was stated above, the soul is different from the *feeling of I*, albeit the latter has some properties that we expect the soul to have. Therefore, is there a soul? We do not know, but if there is a soul, it must rest in the *Region of Existence* and is, therefore, eternal and indestructible.

Let us try to comprehend what we mean by the *feeling of I*. Everybody has this feeling; one knows one is a person. To do some musing, let us assume you cut a finger off and discard it. This is a part of you, but do you miss it? Of course, you miss it, but are you less of a person? Of course not. So maybe a finger is not big enough to make a difference. Therefore, let us assume your arm needs to be amputated. You certainly will miss the arm. But are you less of a person? The answer is no. With this

"Gedankenexperiment", we can go on and remove larger and larger parts of the body. Where will it end? When is a person only half a person? The answer is, as long there is a functioning brain, the person will have the *feeling of I*. Obviously, it is the brain that is solely responsible for experiencing the *feeling of I*. The body is only a support system to provide energy for the brain. Therefore, let us see what happens if the brain is partially malfunctioning. To some degree, everybody's brain does that all the time. That is, we forget things, which we used to know and were thus part of us but are gone now. Somehow we are very sure that this has no bearing on the fact that we are a person, and our personality is still complete. Even if we should have a stroke and cannot talk anymore, we know we are still the complete person.

Reading all these, it may dawn on us now that the *feeling of I* is indivisible, meaning we are here (alive) or we are not. Could it be that the *feeling of I* is indestructible like *energy* and *existence*?

We claim that the feeling of I is the image of the soul, and it is the soul that rests in existence and, therefore, is eternal.

Once the brain has been imprinted with this image of the soul, this image is *real* and, therefore, subject to destruction. Of course, the original of this picture still rests in the *Region of Existence.* This original would have to be the eternal soul. Since the soul rests in the *Region of Existence*, it will not leave the body after death. It was never inside the body, but its image in *reality* (the *feeling of I*) was in the body in the form of electrical charges of the memory cells of the brain.

B. The Mechanism for Realization

A difficult question arises now, namely, how is information that resides in the *Region of Existence* imaged (downloaded in computer jargon) into the brain? According to our model of the world, there is a boundary between the *Region of Existence* and the *Region of Reality.* The image of the soul—*the feeling of I*—is endowed with mass (charges in the brain) while the soul (the original) is not and, therefore, resides in the *Region of Existence.* Logically, there should be no way that one of the two can influence the other.

First of all, does this actually happen? Yes, at least it must have happened sometimes in the past. The program for a human to grow from embryo to adulthood is certainly imaged into the genes of sperm and egg. But that is not all. There are other incidents that are still happening.

One such incident is intuition. By intuition, we mean that we come up with an idea without ever searching for such an idea. One could claim that the brain fabricated this idea without us knowing it. That could be, but is not very likely. There are certainly cases when such a new idea is indeed radically new concerning a problem never thought of before.

Furthermore, there are others incidents. Like the women who woke up one morning and had a British accent. Or that woman who received a heart transplant and developed a craving for peanuts and beer. She also had dreams of a man telling her, "You have my heart." She started to investigate and found the identity of the donor. He was a motorcycle rider frequenting a bar, eating peanuts, and drinking beer, who was killed on his way home.

Still more, there are literally thousands of people who remember, under hypnosis, a previous life. Some of them report details of things that happened hundreds of years ago. Others speak a long-dead language.

Sure, these could all be cleverly planned hoaxes. But again there is always a core of truth in such folklore. Therefore, for the time being, we suggest there is a fuzzy boundary between the

Region of Existence and the *Region of Reality*. Ironically, there is a principle in physics that could describe such a situation. However, due to its complexity, we will deal with this in section VIII separately. Here, it will suffice to say that intuitions and similar realization mechanisms can only be transmitted in short bursts and of a limited amount of information. Since this is a cumbersome way of communication, it is reasonable that nature tries to conserve information that is already here as best as it can be done (see also the information virus in section VIII).

Another question deals with the fact whether it is possible to write a computer program that gives the computer the *feeling of I*. One is tempted to deny this immediately. But remember there were articles written that showed why it is impossible to teach a computer to play chess and win. Obviously, the conclusions of these articles were wrong. But taking a closer look into these programs, we will recognize big differences. Whether or not we can give the computer a *feeling of I*, we will never know. We could teach the computer the answer to all questions it may be asked by persons (psychologists). We would need to teach it also everything about love, hate, danger, and victory. Maybe this could be done, but still this would not empower it to deal with a new situation that was not taught to it. Admittedly, a person has also a problem when being confronted with a new situation. Most of the time, a person deals then with such a situation via emotions. Emotions, almost certainly, may be guided by intuitions. And this is the reason why the *feeling of I* is specific to life, and it is not possible to give a computer the *feeling of I*.

In summary, in this section of the book, we made arguments that, according to our model of the world, the following statements are conceivably true:

1. If there is a soul, it has no matter (energy) and no shape, is eternal and indestructible.

2. The *feeling of I* is an image of the soul. The *feeling of I* rests in the brain and is, therefore, destructible along with the brain.

Mephisto: Did you not hear about the two doctors who weighed dying persons before and after death had occurred? They found out that the soul weighs two pounds.

AT: Yes, I heard about it, and I think to do such an experiment is certainly gross; but nevertheless, there are other natural reasons why a body could lose weight after death.

Mephisto: But you also claim that the soul is not inside the body. All religions so far assumed that the soul is inside the body. European folklore believes that the soul is expelled from the body with the last breath. Other religions claim that it escapes through the nostrils.

AT: Humankind is well aware that somebody who is alive needs to breathe. Therefore, it is only logical that if the soul were inside the body, it would escape with the last breath whether this last breath is discharged through the mouth or through the nostrils.

Ceterum censeo, it does not matter what AT's opinion is. What matters is that in this book, you will find explanations of scientific teachings that will help you form your own opinion.

Section VIII

The Transfer Mechanisms

A. What are transfer mechanisms?

The *Region of Reality* is filled with *energy/mass* and organized by the ordering schemes of 3-D space and time. The *Region of Existence* is filled with *information*. This region has a less rigorous ordering scheme and contains neither mass nor time. If there should be a soul, it obviously must be located in the *Region of Existence* because the soul has no mass. An image of the information, "soul" is located in form of the *feeling of I* in the *Region of Reality.* Nobody can argue that there is no *feeling of I.* Whether or not there is a soul remains to be a matter of belief.

For the sake of answering the subject question of this book, it is extremely important whether or not information can influence energy/mass. If the *Region of Existence* contains the soul and if the *feeling of I* is an image of the soul, this *feeling of I* must be resting in the *Region of Reality*, namely, in somebody's brain. If this is the case, then certainly, there needs to be a communication between the two. Only if such a communication exists can there be a meaningful life after death. The *feeling of I* certainly will be destroyed after the brain ceases to function. Therefore, if the master information (soul) is still in the *Region of Existence*, that situation could be the basis of a life after death.

Anecdotal reports of people who supposedly remember a previous life abound by the thousands. If these memories contain historical events, historical records could verify them. Some indeed turn out to be correct. However, the person reporting these memories could have learned these historical facts by accident. Even if that is not the case, and it is a true remembrance, it does not mean that the person lived before. Such persons could have a gift that enables them to "load down" information from the *Region of Existence.* Therefore, they would "remember" the memories of other people, who are now long dead but did experience the information in question while they were alive.

Certainly, there might be "natural" explanations for all these claims. Logically, they could be verified and explained naturally by scientific research. Such a research team should contain—next to the scientist—a magician, a detective, and a historian. Unfortunately so far, that is not being done.

Let us assume there is communication between the *Region of Existence* and the *Region of Reality*. What gives us the right to assume this? The fact is that there are indeed *intuitions* that seem to come out of nowhere. Everybody has them once in a while. If we are to take people remembering a previous life seriously, then we have to conclude there must be a communication's link between *existence* and *reality.* These intuitions must be information that was resting in *existence* and was picked up by the person having these intuitions. As a matter of fact, almost everybody must have this "gift."

Now, how can *existence* having no mass/energy impregnate the nerve cells in somebody's brain? Logically, there must be some, albeit tenuous, energy positioned on each side of the interface between *existence* and *reality.*

If we were to claim that information (*existence*) could manipulate energy (mass) in the brain, we need to support this with scientific facts.

B. The Uncertainty Relation

The approach needs to be to assume that there is no sharp border between *existence* and *reality*. The border may be fuzzy, and the amount of energy available to spill over from *reality into existence* may be very small.

There is a law of physics known as the *uncertainty relation* that might be able to handle such a situation. Heisenberg formulated this relation in the 1920s. It has two different forms, namely:

$$\Delta E * \Delta t = h$$
$$\Delta p * \Delta x = h$$

(The "*" sign indicates multiplication.)
h: Planck's constant, *E*: energy, *t*: time, *x*: coordinate of location in space, p = m**v*, (*m*: mass, *v*: velocity)

The symbol Δ usually indicates a difference—any difference—but due to the fact that *h* is a very small number, the entities *E*, *t*, *p*, and *x* also must be small quantities for Δ to be meaningful.

The German word *"Unbestimmtheit"* was originally used as a name for these equations. This was incorrectly translated as "uncertainty" into English by somebody, who did not know any better. After several generations of graduate students teaching introductory physics, who did not know any better either, the English word was taken literally. This means it is (erroneously) now believed that this equation has something to do with uncertainty, which means "not quite known." For a scientist or engineer, this implies energy or time or location was measured but not precisely enough. Both are, therefore, known only with some uncertainty, namely, the error of measurement. It is claimed that a more precise determination would change the entity to be measured by the measuring process.

The fact is, however, that these equations do not mean that one has not measured precisely enough or even that one in principle cannot measure precisely enough. The latter is certainly true, but the original formulation of the equation had nothing to do with measurement. Consequently, this interpretation is too narrow. This broader meaning is that these quantities are not *defined* within the boundaries given by this equation.

Now it should be remembered that each measurement involves an exchange of energy in form of a "measurement energy." If the entity to be measured happens to be energy, this "measurement energy" should, of course, be small compared to the energy to be measured; otherwise, the measurement influences the value to be measured. Now going to extremes and measuring extreme small values of energy, it becomes reasonable that such a relation as described by the *uncertainty relation* actually exists. Remember we mentioned Planck's constant h before, namely, in connection with the smallest energy package that can be exchanged in an *energy conversion* involving radiation ($E = h\nu$).

This is simple enough, and all would be fine. However, we claim here that what the *uncertainty relation* describes goes somewhat further. The entity Δ is not "error in measurement," but it is "lack of existence" as the German word implies (here, existence is not in italics; "not defined" could also be used). What is meant is that the value of the distance x does not change over the interval Δx.

To put it crudely, if Δx were to be one inch (an unreasonably large value, here used only to make an explanation easier) and one would put a little ball 1/10 inch in diameter in this region being one inch long, then this ball could be in the location 1.1" as well as in 1.9" without becoming any larger. Subdivisions or subspaces of this region do not exist (again no italics).

For the purpose of using the uncertainty relation for the boundary between the *Region of Existence* and the *Region of Reality*, the form $\Delta E * \Delta t = h$ needs to be used. If indeed there is a very small amount of energy (ΔE) spilled over from *reality* into *existence* at a certain time, then if this time (Δt) is short enough, the small amount of energy can become an entity that is large enough to transfer information into the brain (during this short time). This "entity" is *power*, namely, energy divided by time. The smaller the Δt, the larger the power becomes that is used for the transmission of information.

The *uncertainty relation* is difficult to understand; therefore, try another example. Assume we drop a ball from a tower. It will fall down and bounce back up to approximately the height it started from. In other words, first it had Potential Energy that got converted into Kinetic Energy when falling down. Upon bouncing from the ground, the Kinetic Energy is converted back into Potential Energy until all Kinetic Energy is converted into Potential Energy. At this point, the ball would hang suspended in free air. Of course, that cannot be, not even for a nanosecond. There cannot be a delay between the conversion of all Kinetic Energy and the beginning of the action of the so-attained Potential Energy. It cannot just hang there. On the other hand, it does turn around that implies a stop.

The problem can be solved by applying the *uncertainty relation.* There is a point in time when the atom in the center of gravity of the ball has both—Kinetic Energy and Potential Energy. This means there is a small time interval Δt during which time is identical between beginning and end. There is also a small length interval Δx where the position between the beginning and the end of this interval is identical.

One could also say if the ball is in a position when it should fall according to the fact that its Potential Energy (due to gravity) demands it, then there should be a time interval Δt during which it is not determined whether the ball is moving or not. The point that needs to be made here is that the central atom inside the ball for a given Δx is not at a particular location (x); it rather is all over the range Δx and undefined where it actually is. This accounts for the fuzziness.

C. Special Transfer Mechanism

Existence contains no energy. If it were endowed with energy, we could measure this energy and automatically would also obtain a location in space and time for *existence*. However, we agreed at the very beginning of this book that *existence* is timeless and not located anywhere in space.

For a person to experience an *intuition* requires that information be transferred out of *existence* into *reality*, namely, the brain. We try to solve this problem by suggesting there is a fuzzy border between *existence* and *reality* and citing the *uncertainty relation* as a description of such fuzziness. The uncertainty relation is used in other applications in physics, and therefore, it is not unreasonable to use it here.

In addition to Δt being small, also Δx is very small. As pointed out above, ΔE may obtain a somewhat larger value provided that Δt happens to be small. In this case, the value of ΔE could be sufficiently large to affect something in the *real* world. The smallest energy requirement to afford a redistribution of mass in the *real* world can be found in the brain. Only a small redistribution of electrical charges (by movement of an electron) is necessary to accept information coming from *existence*. Such a movement also has to be physically large enough for it to be sensed. Therefore, there is also a minimum size of Δx required. Consequently, such a transmission has to come in short bursts since the energy to be transmitted is extremely small, but Δx needs also to be significant enough to affect a redistribution of charges in the brain. This is the reason why *intuitions* need to come in short bursts.

This is not such an absurd idea. It is well-known that naturally occurring small electrostatic charges can induce weird experiences in people. Such electrostatic charges can occur in form of piezoelectricity when granite is compressed. This is a possible explanation why certain animals seem to be able to predict earthquakes. Also, some people explain stories of alien abductions this way.

Suppose all these are correct, what are the consequences? Certainly, whatever energy leaks over the boundary from *reality to existence* must be very small, and the duration of the leakage will be short. Now this minute energy must be optimized to get up to a value that can accomplish some changes in the electrostatic landscape of the brain. According to the *uncertainty relation*, Δt needs to be very small for the entity "energy divided by time" (power) to influence a situation in *reality*. Obviously, there must be limitations since Δx cannot be smaller than the size of a receptor in the brain that can sense a small change. Δt may not be long enough for the system to be able to follow and, therefore, to know that such a transfer has taken place. But even if it is too fast for us to recognize that it happened, the intuition is still received. But since we are not aware that it happened, intuitions seem to come out of nowhere.

However outlandish this might seem, it might be that this is a built-in protection system to warn an individual of imminent danger. There were cases when people already seated in an airplane left it without any good reason, only on an *intuition*. And it so happened that this plane crashed. Folklore has the concept of the guardian angel. In most cases, there is a core of truth in what folklore claims.

D. The Regular Transfer Mechanisms

The regular transfer mechanism accomplishes the transfer of information out of *existence* into reality (into our brain) without applying *intuition*. At conception, the new human being is already endowed with an enormous amount of information: the genetic code. The computer jargon for this is "hardwired." However, during the lifetime of a person, other information ends up in the brain as well. This is accomplished by teaching, reading, imitation, and the information virus. Since all information resides in *existence*, this must have been transferred out of *existence* into the brain at some time. Above, we used the metaphor "imaging" for this. When did this happen? It happened as long as humankind is around.

We expressed the belief that *existence* is the totality of information that there is. During our lifetime, a small section of this totality is imaged into our brain. We claim that there are several mechanisms how this is accomplished. One of them is *intuition* that is called above the special transfer mechanism. However, there are others that we call here the *regular transfer mechanisms*. This is teaching, reading, imitation, and the information virus. These mechanisms use what is already transferred to other individuals. For the regular transfer, we must notice that reading, learning, and imitation do not require an energy input coming from *existence*. By energy input, we mean that energy that is required to charge and fire the correct brain cells. In the above cases, we supply the energy required for this; it does not come from an outside energy source. Consequently, we can acquire with these mechanisms only information that already resides in *reality* but not information that exists in *existence* and needs to be transferred into *reality*.

E. Information Virus

Teaching, reading, and imitation are obvious transfer mechanisms. The information virus needs some more explanation. When humankind graduated to electronic computing and later on (via expert systems) to electronic thinking, they improved their understanding of the working of the human brain considerably. On the other side of the coin, nasty people used this knowledge to create viruses and worse and learned how to feed those into other people's computer programs. It is possible that the human computer (the brain) can also be affected by such types of viruses. In this case, we call them information viruses. Why are these relevant for a book on life after death?

Ask yourself what happens when the person next to you reads a book and becomes very emotional, even begins to cry. If you look at this book, nothing happens to you since there are only scrawls on the paper. The book happens to be printed in arabic. Is it not very strange that a person gets emotional just by looking at some scrawls? Could, therefore, such a thing happen just by looking at just a few pebbles scattered on the beach? Based on the above reasoning, the answer must be affirmative. There could be a message resting in the scatter of the pebbles. If we "read" such a contained message, then we were exposed to an *information virus*.

As claimed above, there is limited communication between the *Region of Existence* and the *Region of Reality* in the form of intuitions. This is possible due to the fuzzy border between these two regions. Obviously, nature has come up with other means to transfer information between these two regions. One major transfer must have been the establishment of the genetic code. This is indeed a very large body of information. Official science (Darwinism) teaches that there was no such transfer. The genetic code was formed by random mutations and is improved by selection of the viable mutation by competition. In this book, it is our policy not to question the result of official science because this book would not be the suitable platform for doing this.

Described in this section is one mechanism allowing a person to leave information behind after his death, namely, in form of the information virus. The fact that letter arrangements, Morse code, and other assemblies of objects can contain information suggests that a person may unwillingly leave information behind just by moving objects around. This we call the information virus. The fact that humankind wants museums is a strong indicator for function of the information virus. Other indicators are the fact that persons can get affected by music or just by drumbeats. In the case of paintings, why is it conceived by most of humankind that a picture of a smirking lady (Mona Lisa) has such an immense value? Also, persons like to sleep in a bedroom in which a long-departed famous person has slept before. People touch only reluctantly the block that was used to behead a famous person. And so on.

It is also probable that most everybody can "read" the information virus, albeit to various degrees of sophistication (see also appendix on ghosts).

F. The Points This Section Is Trying to Make

The *Region of Existence* contains information of everything that ever happened and everything that can happen according to the laws of nature. This contains the information that you used during your lifetime. Therefore, it is still all there after you die. If that could be downloaded by another individual, would this person go through similar joys and sorrows as you did? A complete "download" is not likely; maybe fractions of it could be "downloaded" but not all of it. To be reborn in another form of life is not what the Christian religion teaches. According to its teachings, the soul must be staying in a region similar to the *Region of Existence* forever. This region does not contain *time*.

What will the soul do when it is in the *Region of Existence*? Will it experience joy, love, and sorrow? Without having access to energy/mass, this would be meaningless. However, the soul can influence life on Earth by *intuitions*. When doing so, is there a certain selection of intuition to be transferred?

If so, a metaphor might shed some light on this. An entity (here the soul) plays the piano and enjoys the beautiful music. The piano (humankind) does not. It only feels being hammered on and its string being twisted. It has no semantic levels to enjoy the music.

> Mephisto: The ball hanging in the air for no time is news to me. Is this again something new you try to sneak in without being part of accepted science?
>
> AT: It is actually not new. The ancient Greeks already questioned how an arrow could take off from rest. There must be a time when it is both in rest and already moving. Of course, they did not have yet the uncertainty relation to explain it.
>
> Mephisto: Is it not so that a field will spread with the velocity of light? Therefore, your ball could be hanging freely there at least a short time until the gravitational field can get hold of it.
>
> AT: Your argument neglects the fact that the field is already there. The ball is subjected to the gravitational force all the time it moves up.

Section IX

Death and Dying

A. What Is Living?

From the depth of our heart, we believe living means to function in the world of *reality*, not merely to *exist*. Our life consists of events (*energy conversions*); if nothing happens, we do not "live." We feel we need to have control over energy, meaning to be able to manipulate energy; otherwise, we do not live.

We claim that the mission of humankind is to increase the order in the world (see appendix 6 on entropy). To increase the order of the world, energy needs to be manipulated. Whatever humans do causes an *energy conversion.* Humans have no choice.

> Mephisto: How does humankind change the order in the world?
>
> AT: If there is a bunch of stones lying around in the landscape and if humans take these stones and build a shelter of it, the order in the world has been increased.

Mephisto: Why?

AT: Because there is no way that by chance (e.g., during an earthquake) these stones would assemble themselves into a house. If they just lie around randomly, this new state of alignment, namely, in form of a house, would be a higher state of order.

Humankind could be regarded as an intelligent robot—if it were not for the *feeling of I.* Consequently, for the following discussions, we will dissect humankind into two parts, the intelligent robot—the secular human—and the spiritual human, who is endowed with the *feeling of I.*

The secular human is programmed to do certain things that shehe cannot help but do. The most important "instinct" is survival; the next important one is procreation. As we pointed out already, it is the function of the body to provide the brain with energy. The part of the brain that is in charge of the secular functions needs to see to it that the body will continue to be around and so will see to it that food is provided for at all times. Also, since the machine "secular human" will wear out before the mission of humankind can be accomplished, it needs to see to it that this machine is replaced in time with a new one and that this new machine is programmed properly so that it can function in the world of *reality*. All the above-described functions could be programmed into a machine without a necessity for a personality (a *feeling of I*).

Recognizing this, and observing what we are doing, we will be shocked how much activity in our daily life is devoted to a behavior that an intelligent robot could perform. Survival means automatically ducking if an object is about to hit us over the head. Procreation leads to the fight for the desired female between males and the maneuvering of females to hook the desired male. In our younger years, this occupied a major part of our waking hours. In a more mature age, the interest goes more toward caring for grandchildren. Did you ever wonder why grandparents are more devoted to their grandchildren than they were as parents toward their own children? The answer is that continuance of information is to nature a paramount desire. The link grandparent-grandchild is a shortcut for propagating information.

Understanding all these, is there then indeed a need for the spiritual human? Or is there indeed such a spiritual human? In principle, the total operation—self-preservation, procreation, and information transfer—could be accomplished with an intelligent robot only. This might be so, but there is more. Obviously, there is a *feeling of I*, but aside from this, we will show that the spiritual human is indeed required for the mission of humankind (see appendix 6 on entropy).

B. How We End Life

It was already stated that the soul is eternal (if there is a soul). However, what happens to the *feeling of I*, the image of the soul? Does it also exist forever or does it disappear at death?

We speculate here that an individual's *feeling of I* is hacked out of the total *feeling of I* of humanity. Therefore, the least we can say is that as long as life as such is around, somebody will be in possession of a *feeling of I*.

A metaphor may help provide a better understanding of the above statement. Such a metaphor may be a flame; this is probably the closest model of life. A flame needs oxygen, needs food (fuel), tries to grow, and propagates (by spewing out fiery small objects). It even takes action to survive. If one sticks a solid object into the flame body, it attacks this body by the only defense it has, namely, heat. If not successful, it tries to get around it by moving sideways.

We can regard the soul of humanity as a total being, a large flame with millions of flamelets (the *feelings of I*) sticking out of its body. Does such an individual flamelet return back into the flame when it disappears, or is it just extinguished, leaving no trace behind?

In case of the *feelings of I*, at least some traces are left behind, namely, the part of information that was *realized* by our brain but was recorded in one form or another outside the brain. That could be the material we have written, the painting we have painted, the structures we have built, and so on. As such, this information is not active anymore, but it was part of our *feeling of I*. It could be that the *feeling of I* of another human picks it up, and so it would be active again. By "picking it up," we mean a different person reads or sees what we have done, understands it, and agrees with it, and so it becomes part of this different person's personality.

We claim that the feeling of I forms gradually and grows during our entire life becoming more complex with every new piece of information we incorporated. Consequently, we claim also that information we had incorporated into our feeling of I now lives on in those different persons that have seen what we have done.

This is not such an unusual thought; just consider how many Roman or Greek words we use still today (capitol, senator, etc.) and how many ideas of government, philosophy, etc. of ancient origin we still subscribe to. We still carry a part of the *feeling of I* of Plato, etc., in us. Also, the Christian religion teaches that a part of Christ resides in every person who "accepts" him. (I am the vine, and you are the leaves.)

While the fact that we leave something behind may be a comforting thought, this is not what we had in mind by "life after death." Therefore, in the following, we will explore what we can say about the experience of dying and, using reasonable speculation, extrapolate this experience up to and maybe past the actual point of death. Whether or not if this reasonable speculation is indeed reasonable, we never will know since the dead do not come back to tell the living what is beyond. But we also never will know for sure if the universe actually was created by the big bang since nobody was there to observe and tell about it. However, in case of the big bang, we accept circumstantial evidence and conclude from this what might have been. Similarly, concerning the experience of dying, we conclude from circumstantial evidence what might be.

Before we discuss the actual experience of dying, let us ask the question "what dies?" We tacitly assumed already the answer: it is the brain. To make this more plausible, let us ask a few simple questions. E.g., would our *feeling of I* change if one leg were to be amputated? Would we become a different person? We know the answer is no. Even if the heart is removed and replaced by a pump, our personality does not change. We are still the same person with the same feelings. To take it to the extreme, one could indeed remove the whole body, if it were technically feasible to supply blood containing the right balance of chemicals and oxygen to the head. While we hope that this never will be done, we know nevertheless that the personality residing in this head would still be feeling "I."

Therefore, it is a reasonable assumption that the majority of the information, which we expect to be lost by an individual's death, resides in the brain. There are some bits and pieces of information, which are "hardwired" in the nervous system. These are usually called "reflexes." Also, folklore has it that one loves with the heart, angers with the stomach, and frightens with the neck. However, the majority of the *feeling of I* is the "software" residing in the brain (software being everything that is learned).

From where can we get an indication concerning what a person experiences when shehe is dying? Falling asleep would be closely related to it especially all the symptoms we experience when we are fighting to fall asleep (e.g., when driving) should apply. Sleeping means that a part of the system is being shut down (in a certain order) while dying means that the total system is shut down. It is reasonable to assume that this shutdown process (dying) also is done in a certain order to provide (as long as possible) for the possibility of a restart should the external situation change.

Another source of data is the reports of people who almost were there (were "clinically" dead) but were revived on time. Books have been written on these reports. Also, people, who were in a coma and pulled out of it, are a source of information that is available in the literature. There is also the phenomenon of shock, if one wants to make a distinction between shock and coma. It seems that the shock is initiated internally by some mechanism, which is supposed to eliminate pain, when pain no longer can serve as a warning device. For example, an animal may go into shock shortly before a predator eats it.

Therefore, let us do a reasonable speculation on the experience of dying based on the above-mentioned data. Starting with sleep, we know we cannot recall falling asleep when we wake up the next morning. This is different when fighting sleep. When people get older, they do this a lot, even during daytime, mainly for the reason that there is at present no opportunity to sleep or they just do not want to sleep at this particular time. In such a situation, one experiences first some impairment of vision. The system wants to shut down the visual input first and foremost. If the eyelids are not allowed to close, the eye itself seems to fail in its ability to synchronize or rather stabilize the image.

If the brain is to be shut down partially (sleep) or totally (death) in steps, then it is only logical that the main consumers of the brain's electrical signals (energy), which are not required for vital body functions (e.g., heartbeat), are shut down first. One of these main consumers is the visual system. It is of considerable complexity and requires a sizable part of the brain to function and is, therefore, first in the shutdown steps. The peripheral vision is shut down initially resulting in a sort of tunnel vision (pilots call this the grayout). At a later stage, one no longer sees an image as such at the end of this tunnel. The output of the individual pixels are no longer processed and comprehended as an image; only the fact that light intensity is present is perceived. One "sees" light at the end of the tunnel. It is reported that sometimes the image of deceased relatives is seen at the end of this tunnel. In lieu of a physical object for an image, the brain puts an image there from its stored information.

> Mephisto: The claim that the visual system can see things that are not actually there is hard to swallow. Do you have some proof that this actually can happen?
>
> AT: Yes, this actually happens. There was a sailor who crossed the Atlantic Ocean in an extremely small sailboat. Meaning, he was alone a long time and had nothing much to do. A medieval sailor frequently visited him, and they even talked to each other. After the trip was over, he mentioned this in an article and said that he was well aware that this was an illusion but was glad anyway that he had somebody to talk to. Yes, the brain makes up things when there is nothing exciting to see but an empty ocean.

Like an electronic computer, the brain is controlled by an executive routine. It decides which parts of the memory are addressed and manipulated. "Thinking" is manipulation (or reorganization) of the contents of the memory banks. Some large electronic computers are operated by a time-sharing executive (software). That means different tasks are worked on almost simultaneously, meaning the computer works on one task for a millisecond, then stores the obtained (partial) result, then works on the next task for a short time (usually based on a priority scheme), etc. When it returns to the original task, it pulls the partially completed results out of storage, works some more on it, puts it back in storage, and goes again to the next task, etc.

The brain operates differently. It is too slow to accomplish very much in a millisecond. Therefore, its executive routine works on all tasks simultaneously. That means incoming data, e.g., from the eye, are continuously processed on a routine basis and are temporarily stored (so to speak, in a buffer). The executive routine works on these buffer memories and draws conclusion. This it can do only one at a time. This executive routine is perceived as consciousness. Subordinate to this is the subconsciousness, which is the content of the buffer memories. There must be mechanisms that attract the executive routine to the content of a certain buffer memory (e.g., a loud bang will attract our attention to the hearing system, and we listen for more to come). If no such attraction of attention takes place, the executive routine works on a problem of its own choice (not necessarily based on sensory input). "We think." The decision on what to think about is made based on the external situation we are in, and if this demands no action, the logic executive spontaneously picks a subject to work on. Again, the subject is not worked on very long (attention span); after a short while, a different subject is chosen, mostly initiated by sensory input from the outside. If no such input comes in, the mind picks spontaneously another subject (the mind wanders). This wandering of the mind is important. It causes all senses to be continuously evaluated for danger signals even though they do not report any unusual activity. If the sensory input is switched off, in order to save energy (sleeping), the mind quits wandering and works lazily on one subject only. This is dreaming. The level of work for the mind is controllable; it can work lazily on a problem or it can work forcefully on it. We perceive that as different states of excitation.

If we are going to sleep voluntarily, our excitation level is low. The sensory input is switched off, and the mind gets stuck accidentally on one subject on which it works lazily. This subject more often than not is something the mind worked on before or has revisited during the "wandering" process when sleep set in. The initiation of sleep is, therefore, recognized when the mind quits wandering and gets stuck with one problem.

The brain is a large consumer of energy. While thinking, chemical reactions are taking place. Therefore, waste products build up that have to be removed by the blood stream. This process takes time; therefore, it is necessary to restrict the energy consumption for a while, by reducing the thinking activity, until the supply system has caught up. That is the reason for sleep. If the energy consumption and waste product deposition would be continued indefinitely, the brain would just quit functioning; its activity would catastrophically collapse. Intuitively we know this. That is why we panic instead of rejoice when we cannot fall asleep. Or that is why it is a severe torture if we are artificially kept from sleeping. Also, if the sensory input is high, if it constantly demands attention from the executive routine, the latter one will not quit wandering, and sleep will not set in. (We cannot sleep in a noisy brightly lit environment.)

In dying, the situation is similar. The brain tries to shut down, at least partially, by shutting down the larger energy consumers (e.g., vision). The reason may be illness or exhaustion. The symptoms of shutdown, as pointed out up above, are tunnel vision with a light at the end of the tunnel. In case the mind wants to fight the shutdown, this is done by increasing the excitation level, and instead of slowing down its wandering, the mind begins to race. This results in a recall of almost all our life experience (famous tale of people who fell out of windows and lived to tell). Panic sets in, and the memory banks are now searched for help. Such can usually be expected from close relatives such as father, mother, etc. Therefore, although the visual system is shut down to a degree that only a tunnel with a diffuse light at the end is perceived, these close relatives are seen clearly and are recognized. They are put there beyond the diffuse light by the memory, rather than by the visual system. This can go on for an extended time until all energy stored in the brain cells is consumed and the blood no longer replacing this energy.

However, at this point, we may be dead already. The functioning of the brain as a retriever of memory content is not sufficient to qualify as being alive; otherwise, one would have to regard a computer as being alive as well.

We are alive as long as we have freedom of choice and being subject to *intuition*. Only if we have a choice on what we want our mind to work on are we conscious, which we consider (in this book) as being alive. That means we are already dead (in a simulated death) each time we are sleeping even if we are dreaming; only it is not permanent since we entered this situation voluntarily. Therefore, in this case, none of the panic searching has taken place.

It is entirely feasible and reasonable to assume that a person who wants to die could enter death the same way as falling asleep, without any panic searching, even without tunnel vision, since the latter takes place only if we fight sleep.

It is also entirely feasible that a person who does not want to die can keep himself alive by sheer willpower. The mind will keep searching for things to work on, even though the choices may be less and less. If there is not enough energy supplied, most of the brain will be shut down, and the mind has only a few problems left to kick around. Since all sensory functions are shut off, there is no communication to the outside, and the person is said to be in a coma. From this, we can see that dying can also be very slow. The rest of the body may already be dead and below body temperature. Somewhere inside the brain, there may still be some energy left, and "life" will go on as long as this is the case. There were cases where people were brought back to life from such a condition.

We should not really be sorry to die. But shouldn't we really be fighting death? If we have the energy left, yes, we should. However, when we lose, there is no reason to be sad. *The body is a disposable container for the feeling of I, which is the image of the soul resting in the Region of Existence.*

Death is, therefore, a reassignment of the residences of a (small) part of the *existing information*. It is not that a part of the soul is destroyed when our brain ceases to contain the *information* it used to contain. Since the soul does not consist of energy, a physical presence—and, therefore, a defined location—of the soul in space and time is not possible. Only energy is present (*real*) in the physical world. Since the soul is the sum of all information residing in the *Region of Existence*, it is not destroyed if the residence of some of the (imaged) information is changed. Any information residing in *existence* cannot be destroyed since it is not a physical entity. The structure of matter can be changed—by a process that may be a destruction, like the decay of the brain tissue. However, the *information* itself that was in the brain cannot be destroyed. The thoughts we thought still *exist* after our death and always will. If we communicated them, they might be continued by other brains; if we did not communicate them, they might be rediscovered and reexperienced by other brains.

The total sum of all our thoughts was our "I." The thoughts that are continued by other brains, therefore, contain a certain small fraction of our "I."

To recapitulate this:

1. No, we are not waking up after dying somewhere else remembering all the things we did during life on Earth.
2. Yes, there is continued existence. "I" will exist as long as there are containers (brains) for it. Our "I" is not different from other "Is." This should mean that after death, even if there is no memory left of our life on Earth, "I" still exists and continues to exist because other people will feel an "I." In the last analysis, this means that all other people on Earth are just different sides of our own self.
3. Having said this, if it indeed happens that gifted people can "load down" memories of a previous life, however, this does not mean that it is their previous life; rather, these may be (partial) memories of a life of somebody else. Remember, the *Region of Existence* does not contain time.

And indeed if one subscribes to the idea that there is a common soul of all humankind, this seems do be a redeeming fact that makes dying more acceptable.

Ceterum censeo, it does not matter what AT's opinion is. What matters is that in this book, you will find explanations of scientific teachings that will help you form your own opinion.

Section X

Life after Death

A. Important Remarks

This is the last section of the book on life after death. If you succumbed to the temptation to read this last section first rather than to labor through the entire book, you may be unable to understand what is said in this section. We implore you to read the book first before you read this last section. The answer to a profound question such as asked in this book cannot be obtained without serious effort and labor. The book made an attempt to explain the involved subject matter in understandable terms, but still an honest effort will be required to achieve understanding of what is said. If you are trying to circumvent this, by reading the last chapter first, you are depriving yourself of being successful in achieving an understanding of the subject matter. Even worse, you will not even recognize that you do not understand, and you will not read the rest of the book at all and so missing something that would have been valuable to you, regardless of whether you agree or disagree with what is said in the book proper.

B. The Region of Existence

In our book, we brought forward the idea that there must be a region of the world, which we called the *Region of Existence*. To make sure again, this is not a concept the physical sciences can teach. These sciences restrict themselves to the exploration of the observable region of the world, namely, *reality*. *Existence* is not observable since it is not endowed with energy. Observation can only be achieved if different forms of energy are exchanged during the observation process.

Since the *Region of Existence* cannot be observed, how can we say there is *existence*? We claim that *existence* is obvious, although it cannot be observed. (It is an axiom.) An important point is that *existence* can be described. The very fact that it can be described, namely, that there is logic and semantics (remember, mathematics is applied logic), makes it obvious that there is a *Region of Existence*. Such a region of the world is of unlimited extension (but not infinite) and is filled with information. Therefore, what *energy* is to *reality*, *information* is to *existence*.

Consequently, one would be tempted to say that the claim that there is a world of *existence* is a major breakthrough in philosophical thinking. However, the idea is indeed not new; it is rather very old. Descartes said more than three hundred years ago, "Cogito, ergo sum" (I think, therefore I am). In all periods of the history of philosophy, one finds bits and pieces of this idea.

C. Is There Life after Death?

Again, this question was already answered in the book proper. Here, we try to summarize what was said. First, what do we mean by the word "life" in the above question? Does this mean we rise after death and keep on doing the things we were doing all along, just in a different world? If this is the question, the answer has to be *no*.

If we ask the question: "Is there continued existence after death?" the answer would be *yes*. Here, the word "existence" is not printed in italics. If it were printed in italics, we indeed would have the closest thing to a proof—so to speak, conviction beyond a reasonable doubt—that this is true. We "know" that information is indestructible; therefore, it will always exist.

We have to admit that we may be accused of having answered a nonquestion. What a person really wants to know is whether shehe will remember her/his life on Earth and whether shehe will be rejoined with his/her loved ones. Therefore, we will now try to answer these questions consistently with what was said in the book.

It was stated that the *feeling of I* is a slice out of an all-encompassing common soul of the entire humankind. The individual *feeling of I* is an image of a (small) fraction of the all-encompassing soul. By "entire humankind," we mean all the people who ever lived and ever will live.

We used the metaphor of a flame. This big flame represents the soul of the entire humankind. From this big flame, many flamelets emanate. These flamelets are the souls of individuals. The *feeling of I* of an individual person is an image of one of these flamelets. The information contained in such a flamelet *exists* in the *world of existence* and was transported (*realized*) into the brain by various transfer mechanisms. After such a transfer, this information is hosted by energy/matter, namely, the brain. Only in this condition can *information* manipulate *energy* and increase the order in the world. We claimed that the mission of life is to increase the order of the world.

As pointed out in the appendix on entropy, the order in the world can only be increased (or the entropy decreased) if energy is manipulated. In the same appendix, it is also pointed out that entropy is not a substance, as *energy* is; entropy rather is information.

Now the question is what happens when the host energy/matter—meaning here, the brain—disintegrates? In this case, the complex organic molecules disintegrate into simpler molecules, no longer carrying the *information* content of the original organic molecule. Is this information, hosted by these organic molecules, lost? The answer is obviously affirmative. Yet we claimed before that information cannot be destroyed. This is still true. The flamelets still *exist*. However, such *information* contained in them resides in *existence*. This *information* was *realized*, meaning imaged onto the brain. This image of information is destroyed at death but not the information itself. This information is still residing in *existence*; maybe it is now *submerged in the large flame proper*.

During all our lifetime, we incorporated more and more information into our brain, all becoming a part of our *feeling of I*. The sum of all these is us (ego). This will certainly disappear.

Actually, some of it starts to disappear even before death. As we start forgetting things, we are losing part of it already. From this, we conclude that losing part of or all the *realized* information cannot be such a grand tragedy. We claim here that the information is still there (in *existence*); it rather is only no longer *realized*. Of course, it could be realized again into another human being, maybe not all of it and not in this particular arrangement. The fact that some people seem to be able to come up with knowledge they did not learn may make us speculate that bits and pieces of a particular arrangement of information bytes stay together and are sometimes *rerealized* in one chunk. By definition, we cannot explore the structure of *existence*, but we could draw conclusions from strange happenings. Established science could do this, by doing some detective work. Yet this happens only to a minor extent.

While the above indications of partial reincarnation may be interesting, but even if correct, such conclusions will be of no practical interest. "Practical interest" here is the question as to if we will still "live" after death. If we mean by "living" the ability to manipulate energy, we would have to claim that the only way to accomplish this would be for the information that made up our ego to be *rerealized*. We cannot possibly hope that all the bytes stay together, in the combination as they were in, and end up in the next person. We could hope, and probably correctly so, that some of our ego ends up *rerealized* again, most likely in our children and grandchildren but also in people who read what we wrote or see what we painted, etc. Therefore, we have to confine ourselves to the recognition that we are part of a larger assembly of information, namely humankind. As long as there is humankind, there will be individuals who enjoy "I" and will be endowed with information bytes some of which were once ours and most importantly are still ours. One could say we are *indeed living on*, albeit *disassembled into different fragments*.

This may sound like a radical view contrary to Christian teachings. A closer look shows that this might not be entirely contrary to Christian teaching.

We may claim that the Christian message comes in different levels. The basic level is that God loves humankind, that God forgives our sins, and that we should love our fellow humans (including our enemies). There are other levels of the message that almost obscure the basic message. These levels were added by the apostles for reasons to gain followers and later by the religious bureaucracy, sometimes for purely self-serving reasons. Nevertheless, the bureaucrats could not ruin the basic message, although they certainly tried hard. This shows the basic message is indeed enormously powerful. For the following, let us stick with the basic message.

Could it mean that Jesus is the personification of all humankind? After all, human beings are God's children. The Christian faithful prays, "Our *father* in heaven." Jesus says (in John 15:5), "I am the vine, and you are the branches." Could this mean that when we die, the bad bytes of information (sins), which we *realized* but should not have *realized*, are separated from us (forgiven)? Hopefully, the next person does not *realize* these bad bytes again. But if shehe does, shehe will be forgiven too. We die for this purpose of purification. "Death is the wages of sin" (Romans 6:23). If Jesus represents the personification of humankind, does this mean, since he was resurrected, there is redemption for humankind?

Here, we are just asking these questions; we are not saying it is so. It is up to each individual to believe what shehe thinks shehe should believe. This book is an expression of an opinion of one individual human being. It is not a proclamation as to how things are because fortunately (or unfortunately) nobody can know for sure how things indeed are.

> Mephisto: In this section, you seemed to pinpoint fairly extensively what happens after death. My boss will not agree with you. Is this all opinion, or can you tell me what you claim you know for sure and what you speculate and where these all come from?
>
> AT: I think so, and I will try:
>
> 1. Yes, the whole book is only my opinion. But I am willing to tell you how I arrived at this opinion.
> 2. What I am sure of is that we live in a three-dimensional world. This world is organized by the ordering schemes "space" and "time." Yes, the three-dimensional space may be bent into

a fourth dimension. This I do not know. And if it were so, it would have no influence on my opinion on life after death.

3. What I infer (not speculate) is that there is a *Region of Existence*. You cannot claim that there are no ideas, and you cannot claim that these are destructible. How would you destroy an idea? By burning books? By forgetting? Ideas are around as long as there is humankind. Forgotten ideas resurface all the time.
4. If there is a soul (this I do not know, but I speculate that there is one), then this soul must be a part of the *Region of Existence*. As such, it must be eternal and indestructible.

Ceterum censeo, it does not matter what AT's opinion is. What matters is that in this book, you will find explanations of scientific teachings that will help you form your own opinion.

Postscript

Something to Think About

The Mission of Humankind

We claimed in this book that the mission of humankind is to strive to decrease entropy or, in other words, to increase the order in the world. In religious terms, that would mean to fight evil and abet good.

Decrease in entropy can only be achieved by manipulating energy. While energy conversions take place naturally (by itself) ever since the big bang occurred, all these conversions lead to increased entropy.

We claimed in the book that a decrease in entropy could only be accomplished by life. Life evolved until it culminated in humankind. Then the anatomical evolution paused, and cultural (social) evolution proceeded. Grand manipulation in energy and therefore grand reduction in entropy can only be accomplished by humankind rather than by one individual.

Contrary to what cynics would claim, humankind has improved in his morals over the last one hundred thousand years by developing civilization. Through advancement of civilization, humans could master the manipulation of increased amounts of energy. They also succeeded in avoiding exterminating themselves by the use of weapons of mass destruction. The reason is that to accomplish such an act of mass killing would take the cooperation of at least a part of humankind rather than one individual. However, humankind, by developing civilization, has grown into an organism of its own obeying higher laws and, therefore, will resist such an act. A hundred years ago, humans did not hesitate to eliminate a whole race of people (like in case of Tasmania). Today humankind is unwilling even to eliminate a species of animals, let alone a whole race of people. Of course, there always will be individuals who will try to do so and have tried until recently. However, nowadays humankind will step in, albeit belatedly, yet they will.

The Creation of Humans

We claimed in the book that the transition from animal to humans happened when one particular individual did something that did not provide any benefit to this individual. This happened when humans buried their dead. Thus, religion was introduced into humanity. We claimed this happened by an outside interference and not by dictate of the genes. Albeit it might have happened only once, the rest of the individuals in close vicinity would imitate this, and later generations would improve on it. Civilization had started. Families and groups of families existed before, even animal have those. It took this particular feat, namely, to do something (e.g., to love your neighbor) that is not in the individual's interest to create civilization and, therefore, humankind. Religion is a set of laws that allows humans to coexist whereby individuals are induced to obey these laws without being coerced by an outside power to do so.

Death

To fulfill the mission of mankind, millions of individuals and thousands of generations are necessary. For plain physical reasons, the human body can only be maintained for a limited time before it wears out. Therefore, death is necessary. However, that also means that there is a need to conserve the information already *realized* by humankind as a whole. Logically, the information possessed by an individual needs to be conserved as well. If this would not happen, humankind could not progress toward its goal and fulfill their mission. While information residing in *existence* is indestructible, *realized* information is not. However, only *realized* information can enable humans to manipulate energy in a way that entropy is decreased.

Certainly, humankind can retain realized information by teaching the next generation. However, this is an inefficient process, and only a tiny fraction of already realized information could be perpetuated this way. Therefore, since nature is known to be very efficient, one can safely assume that *realized* information is also transmitted by other means. It may even be held together in a cluster in *existence* and rerealized by an individual of the next generation. One would, so to speak, continue to exist in another person.

While there is, of course, no proof that this is so, an example may be a person suffering from total amnesia. Shehe lost all herhis memories; shehe does not even know herhis name. This amounts to the same as if the former person that shehe was had died. Shehe would no even recognize herhis mother if she walked in the room. Since this is so, and since this may be a good model for death, how can anybody hope to be reunited with her/his loved ones after death?

However, this example shows that the person prior amnesia lives on has a *feeling of I* and enjoys life. And after all, there is something left of the old personality; shehe can still speak the language that shehe was taught as a child, and other pieces of information also are still left. Does this not amount to the same as if the other person had died and some information (a small cluster of realized information) was transferred to the new person?

The new person is still in the same old body; however, would it make a difference if it were a new body?

Ceterum censeo, it does not matter what AT's opinion is. What matters is that in this book, you will find explanations of scientific teachings that will help you form your own opinion.

Appendices

Please read the appendices. Only after understanding these appendices will you have a true feeling what the book is trying to say. It may not reflect what you do believe, but you were exposed to an alternate point of view.

Appendix 1

Energy

A. General Remarks

Why do we need to learn about energy in a book on life after death? As we have seen in the main sections, energy and time form the *Region of Reality*. Therefore, we need to understand *reality* before we can understand a world where we may be going after we leave the world of *reality*. If we do not perceive noise, we do not appreciate silence.

B. Kinds of Energy

There are different forms of energy. We try to understand these different forms first before we try to explain what energy actually is. Here, we claim there are *three* forms of energy. These are

Potential Energy
Kinetic Energy
Heat

You may have heard about other forms of energy like solar energy, sound energy, deformation energy, internal energy, radiation, etc. Present-day science considers all these as different forms of energy. We claim these forms can all be fitted into our three categories. The reason we are doing this is that we try to understand what energy actually is rather than worry about advanced details.

Therefore, here we go:

Potential Energy is linked to space.
Kinetic Energy is linked to time.
Heat is linked to both, but refers to randomly moving matter.

AT realizes that these definitions could have been invented by the oracle of Delphi. In order to make it more understandable, let us start with *Potential Energy*. Take a stone and take it on top of a tower. Hold it over the rail ready to let it drop. At this moment, the stone has Potential Energy. The minute you let it go, it starts to convert its Potential Energy into Kinetic Energy until it hits the ground.

Why does this happen? Earth is surrounded by a gravitational field. These are lines of force, and a body attracted by the gravitational field of Earth will fall along these lines. In case of Earth, these lines are practically perpendicular to the surface of Earth. This means where a particular gravitational field line intersects with the ground, it will be perpendicular to the ground. This sounds very scientific, but it is nothing else but a more involved description of what happens. It is not an explanation why it happens.

Why do heavy masses attract one another? The answer is we do not know. Sure there are theories about gravitons and gravity waves. So far, these are only theories and not very useful ones. (Theories are never wrong or right; they rather are useful and not so useful.) A more useful one is the theory of relativity. It states that by the mere fact that a mass is present in a certain location in space, the space surrounding this mass is stressed. The jargon for this theory is that the metric of the space is changed. The result is that a connection between the point A and B that was a straight line before the mass was put there is no longer a straight line after the mass was put there. Einstein conceived a nice demonstration as how the surrounding space is stressed by the mere presence of a heavy mass. This is his billiard table. We describe it in appendix 4 in more detail. We stated above that *Potential Energy* is related to space. Now you can see why this claim is made.

But nevertheless, all these are only a description of what happens, not an explanation. There is a school of physics that states that a question as to why this happens is an inappropriate question. The belief of this school is that there must be basic facts that are self-evident like axioms or a-priori probabilities. Only after these facts are accepted can physics extend on it to explain more complex systems. Well, so be it.

A simple way to obtain a certain feel of this interplay between *Kinetic Energy* and *Potential Energy* is to take the example of a pendulum. This may be the pendulum of a grandfather clock or just a string with a weight attached to it. Figure A1 shows such a pendulum. If we move the pendulum into a position as shown in position B and let go, the pendulum will start to swing. A *moving* mass—here represented by the weight of the pendulum—is considered to have *Kinetic Energy* during the time while the mass is moving. That is why we claimed that *Kinetic Energy* is related to time. Of course, we cannot see this energy (we can see the mass moving but not the energy itself). Then how come we know it is there? Well, if we put a piece of glass in its way, the glass gets shattered. It takes energy to do this.

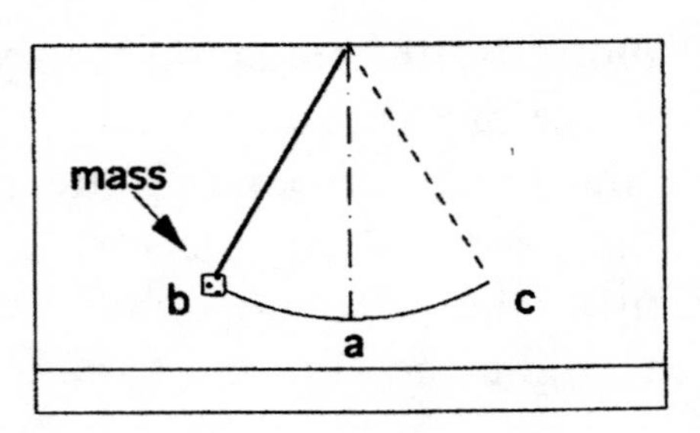

Figure A1 Pendulum

Why do we say it takes energy to shatter a glass? Well, it certainly takes work to do it. Work and energy are scientific synonyms.

Fine, let's go to position C. Here, the pendulum stops for an instant before it reverses its direction and starts to swing again. The fact that it stands still for an instant tells us the pendulum has at this instant no *Kinetic Energy*. But it is farther away from the ground than it was in position A. Therefore, if we were to cut the string in position C, it would hit the ground harder than if we were to cut at position A. Consequently, if the pendulum possesses no *Kinetic Energy* in position C, it must have some other energy that is greater in position C than in A. This other form is called *Potential Energy*. In cases close to the surface of Earth, the distance from the ground can be considered as a measure of *Potential Energy*. We all know it is hard work to lift something. As pointed out above, the word "work" can be substituted for energy. We understand now that Potential Energy is something a stationary body may possess, just on the merit of the fact that it is at a certain position in a gravitational field. As pointed out above, *Potential Energy* is related to space. The name "potential" is unfortunate. This body does not have a *potential* for energy (in the normal sense of the English word); it indeed possesses this form of energy already.

So far, we understand that the weight/mass (for the time being, do not worry about the difference between mass and weight) of the pendulum sometimes has *Kinetic Energy*, sometimes not. Sometimes it has more *Potential Energy*, and sometimes it has less. The next step to understand is that, on above example, the sum of the two energies stays constant. In other words, whenever the Kinetic Energy decreases, the Potential Energy increases and vice versa. Or in still other words, the *Kinetic Energy* is converted into *Potential Energy*, and *Potential Energy* is converted into *Kinetic Energy* as long as the pendulum swings. This leads us to the next phenomenon. After a while, the pendulum ceases to swing. We explain this to be caused by friction. There is friction between the weight and the air; there is also friction at the point where the pendulum is suspended. The friction converts *Kinetic Energy* (and only *Kinetic Energy*) into *Heat*. As pointed out above, *Heat* is the third form of energy. Heat has the tendency to spread itself around. This is done either by heat conduction or thermal radiation. In other words, anything hot eventually cools down by heating its environment. Therefore, heat is an entity that cannot be stored. Its dissipation can be slowed by insulation but not prevented. *Potential Energy* or *Kinetic Energy* does not disappear by being spread out. Of course, *Heat* is not destroyed either; it rather is spread out.

Now it is time to introduce the first law of thermodynamics. Sometimes this law is also called the law of *conservation of energy*. The word "conservation" in connection with energy is unfortunately also misused in another context. This happened when it became time to tell us that we should not waste energy, the bureaucrats did not say so in so many words; they rather came up with a word that sounded bureaucratese. They picked "conservation," although it was already taken for some other meaning. Now what is the meaning of *conservation* of energy? In the original scientific sense, it means that energy can neither be destroyed nor be created. It can only be converted from one form of energy into another different form of energy. In other words, at each energy conversion, the total amount of energy before and after the conversion has to remain constant. This in turn means it is impossible to build a *perpetual motion* machine (*perpetuum mobile* in the European literature). Such a machine would generate energy without requiring fuel at all; or at least it would require a smaller amount of energy to operate than the amount of energy it would produce.

Is there proof for this to be true (the first law of thermodynamics)? No, there is no proof, but so far no violation was encountered, and maybe after you finish reading this appendix, you will agree that the first law indeed makes sense. For the purposes of this book, we will call such an exchange of energy from one form of energy into another one an *energy-conversion event*.

At this point, we like to invent a "fourth" law of thermodynamics. (There is already a second law, and sometimes Nernst's rule is also called the third law.) Our "fourth" law says that not only is it extremely difficult to avoid friction, it is in principle impossible. Therefore, the fourth law states that in any *energy-conversion event*, all three forms of energy participate. This may seem to be nit-picking and not worthwhile to state it in the form of a basic law of physics. However, this law may have far-reaching consequences. So far, we have not told you what energy is. The trouble is energy cannot be defined. It can be described in the form of mathematical equations. (For example: $E_{kin} = \frac{1}{2}mv^2$ for Kinetic Energy; m: mass, v: velocity) However, such is a mathematical description, not an explanation. In other words, this equation is an instruction how to measure energy. To follow this instruction, there is no need to know what energy is. The equation just says measure the value of the velocity and multiply it by itself, then measure the mass and multiply again, then divide by two.

We will try now to explain what energy is. This explanation is the heart of the present appendix. Everything said so far was only said to help you understand this explanation.

We claim that energy is a substance and is the only substance there is.

We did not say a "material"; we said "substance," which is a little more vague. By material, you may envision a piece of steel. A material has properties (color, hardness, density, etc.). By choosing the word "substance," we try to get away from implying too many properties, but we require that there have to be at least one, for a substance. What are these properties? It cannot be color since energy is not visible. The answer is there are only two properties that the substance "energy" has. These are the *mass* and shape. Shape is obtained by the fact that mass occupies space. The quantity of energy is measured in kilowatt hours (see your electricity bill) or in joules. The mass is measured in kilograms (or pounds if you insist).

It was stated that energy is the only substance there is. What about steel, butter, milk, and honey? Are these not substances? This certainly deserves an explanation. Sure, all these items and more are substances. But we know they are all made up of atoms. And the atoms are really only made up of *protons*, *neutrons*, and *electrons*. All other fundamental particles are unstable. What makes the difference between steel and butter is the arrangement and number of these three basic particles. Realizing this, all substances are now reduced to three basic ones. What are the properties of these three basic particles? They certainly do not have a color. Protons and neutrons are believed to have a certain diameter (this implies they are little spheres, which is a naive belief). Electrons have an inconsistent diameter, if at all. Now it becomes fairly tenuous. It boils down to saying the properties of the basic particles are *mass*, *charge*, and *energy*. Charge is represented by the electrostatic field of the nucleus and the electrostatic field of the (free) electrons. Neutrons do not have a charge and are stable only inside the nucleus. Outside the nucleus, they decay into a proton and an electron. In our definition, *charge* or the electrostatic field provides the home for Potential Energy. The field itself is not Potential Energy, but a particle placed into such a field possesses Potential Energy. The same is true for a gravitational field; a mass placed into such a field has Potential Energy, but the field itself is not energy (therefore, do not try to invent *perpetual motion* machines designed to drain energy from the gravitational field of Earth). Do not despair if we are talking above your head, for we take a now a side step to remedy this situation. Therefore, as a side step, we discuss "mass" and then go on with the description of "energy."

C. Mass.

Figure A2 depicts a "balance." It tells the user whether or not two masses are equal.

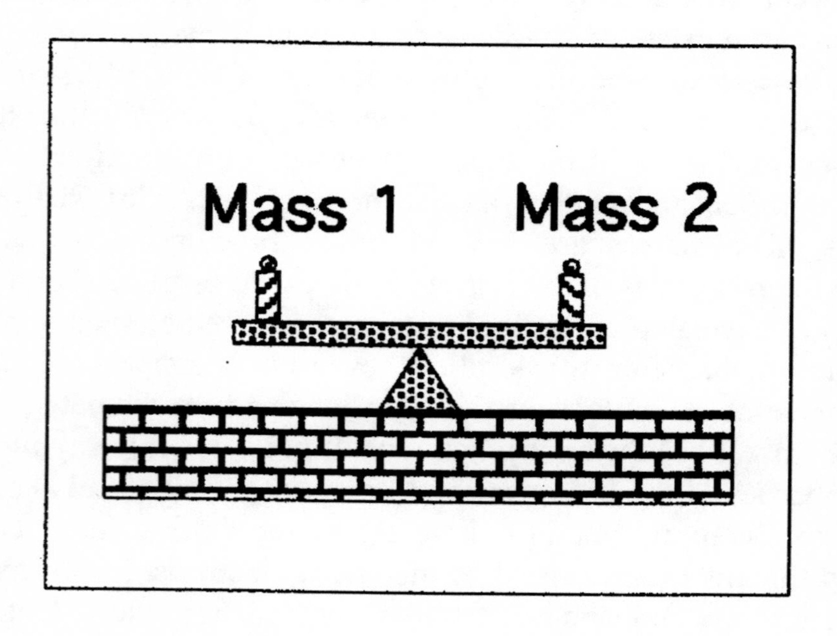

Figure A2: Balance for Comparing of Masses

In figure A3, we show another measuring device that, as it will turn out, is a weight-measuring device. The name is "spring gauge."

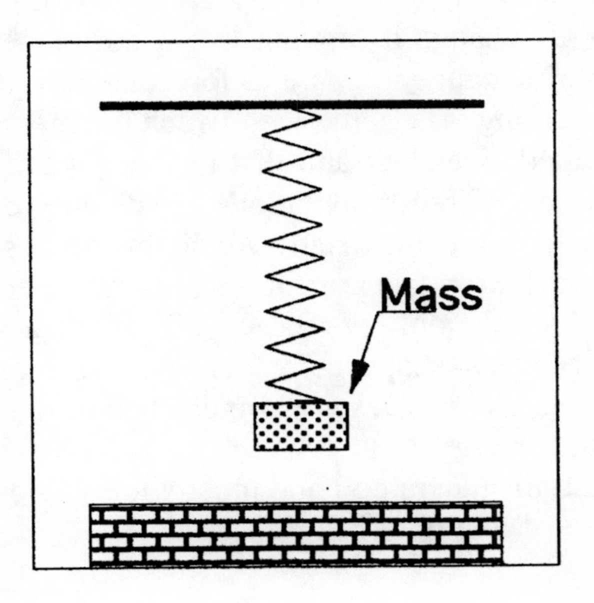

Figure A3: Spring Gauge for Determination of Weight

A spring is pulled by the weight to be measured. The amount of elongation the spring experiences can be calibrated in terms of kilograms or pounds, which are masses. It could be also calibrated in terms of newtons, which would be weight. Of course, nobody is interested in weight, but everybody uses the term, which explains why the difference between mass and weight is so difficult to understand. Let us now perform a *Gedankenexperiment*. Take a kilogram of mass and measure it on the balance seen in figure A2. The result would be one kilogram. Now perform the same measurement with a spring gauge calibrated in terms of mass. Again, the result would be one kilogram. Now go to the Moon and perform the same two measurements. Lo and behold, the results will now be different. The balance still will return the result of one kilogram while the spring gauge will claim it is only 1/6 of a kilogram. Why? The balance in figure A2 compares masses. Masses do not change if brought into a different gravitational field. The spring gauge measures a force, namely, the force that pulls on the spring. The force with which the mass pulls on the spring is called "weight," and that changes when the mass is brought into a different gravitational field. Why do we bother you with such nit-picking detail? We do this because it turns out that is the most important part of the present appendix. "Mass" and "shape" are the only properties of the "substance" energy. Mass itself can have shape, color, density, and so on. From this, the world is made up. Despite popular belief, mass cannot be converted into energy.

The most misunderstood and most widely known equation of the theory of relativity is

$$E = mc^2.$$

(In plain English, Energy [*E*] is equal to mass [*m*], multiplied with the velocity of light *c* and then again multiplied with the velocity of light.) The above equation is called the mass-energy equivalence. It is not called the mass-energy equality. When this equation came out as a consequence of the theory of relativity in the beginning years of the twentieth century, it was written with an "identical" sign:

$$E \equiv mc^2$$

Most mathematical equations need an interpretation. Not this one. If both sides of the identical sign are identical, obviously mass cannot be converted into energy. In less dramatic terms, this means all energy is endowed with mass (because mass is a property of energy).

But what happens in nuclear fission? The nuclear engineers have already coined the term: "mass defect." This is the amount of mass (about 200 MeV) that disappeared during fission. And how is this determined? Take the mass of the uranium atom before fission and compare it to the mass of the sum of the two fission products and other side products. The result is that there are indeed about 200 MeV missing. Where are they? The answer is that fission fragments are moving (have Kinetic Energy) and, therefore, have more mass than when they are at rest (according to the Lorentz transformations. See appendix 3). The masses of the atoms, which are the fission fragment, are known for the case when they are at rest. But when they are moving, their *Kinetic Energy* has a certain (additional) mass. And that is exactly how the 200 MeV were computed. Mass was not converted into energy, rather one form of energy (in this case, the binding energy of the fundamental particle in the nucleus—a *Potential Energy*—was converted into Kinetic Energy).

Sure enough, there are incidents when "mass" completely disappears, for example, electron-positron annihilation. The reaction products are two gamma quanta. Those do not have a rest mass, but they have energy, and any energy is endowed with mass. Things get a little confusing here. First, the positron is an antimatter particle. Second, electromagnetic radiation (gammas) may not have a rest mass, but it certainly has momentum, which normally is linked to mass. The whole reason for this excursion is to provide an understanding that there seems to be two different types of masses, namely, masses that are heavy (*heavy* means that they respond to a gravitational field) and some that have *inertia*, which means that if a mass is moving, it will stay moving unless it is stopped by an object. Established science has two names for it: *heavy mass* and *inertial mass*. Fortunately, Einstein decided that the conversion factor between the two is equal to one. In other words, there are no two different masses. We claim here that mass is a property of the substance energy. How this property reveals itself depends on which environment the mass is placed. If it is put in a gravitational field, it pushes with a *force* on the support that it is resting on (mass is not a force, but weight is). If the mass is placed outside a gravitational field and it is made to move (which is acceleration), it resists such an acceleration with a force (known as Newton's law).

To describe it less complicated, take a rifle and fire it with a certain kind of round. Measure the kick that you are receiving. Now you go to the Moon. Take the same round and weigh it. You will find out that it weighs only 1/6 of what it did weigh on Earth. Now fire the rifle using this round. What kind of kick will you be receiving? Is it 1/6 of what you received on Earth, or is it the same? You do not go to the Moon to get an answer to this question. You get the same strong kick as on Earth.

So what? The important reason we go through all these—for the context of the book—is to help you understand that energy rests in both parts of the *Region of Reality.* There is the "three-dimensional space," and there is "time." This is the world we are living in, and we have to understand that this is not the place we can go to after death. "Heavy" mass is located in the—three dimensional space (namely, in a gravitational field regardless of time) while inertial mass is located in time. (The mass moves regardless of where it is in space.) Any mass there has both *heavy* and *inertial* properties.

In conclusion: A property of the substance "energy" is mass, which extends into both—space and time.

Let us introduce a final comment on the ordering schemes space and time. What do they order? The answer is they are entities that enforce the law of causality. Time sees to it that the *cause* always comes before the *effect* while space sees to it that every *effect* has a *cause*. We apply this to a simple example. Figure A4 shows a simple transmission. Kinetic Energy is fed in on one side (the left) and is taken out, somewhat modified on the other side (the right).

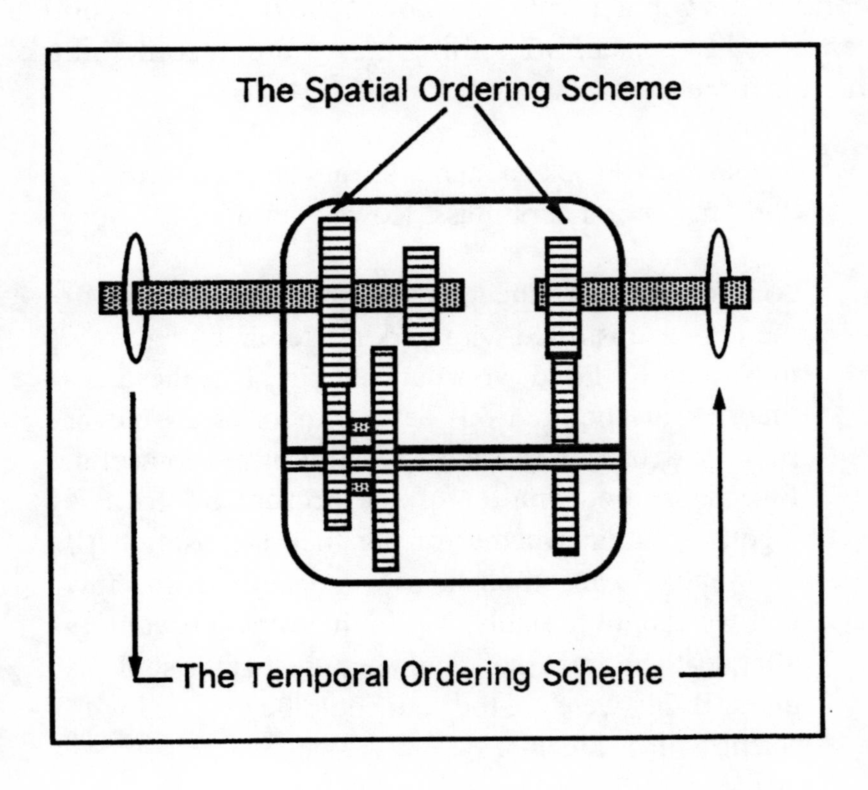

A4: Space and Time as Ordering Schemes

To satisfy the law of conservation of energy, it would be necessary that the appropriate amount of energy comes out at the right side. It could be that this amount could come out before the energy was fed in on the left side. That is obvious to anybody. But why "obvious"? Only because we know by experience that this is so. The reason rather is that *time* requires that the cause comes before the effect. Now it could be that the gears are in a position in which the last gear wheel is not engaged. While energy could indeed be fed in at the left side, there would be no output on the right side. Since energy is a substance, it has to flow through the transmission, which can only happen if there is a continuous contact from one side to the other. This continuity is enforced by space. Something has to touch something else, which is done by space.

> Mephisto: You talk about mass, but you fail to mention dark matter and dark mass. How about this?
>
> AT: Yes, there are theories that proclaim the most of the universe is filled with dark matter and, therefore, must also be filled with dark energy. But these are theories. As pointed out before, theories are never right or wrong; they are only useful or not so useful. For the subject matter of the present book, it is inconsequential whether or not they are useful. All we wanted to accomplish in this appendix is to show that we cannot possibly stay in the *Region of Reality* after death because that would require that the soul has mass. If this were so, the soul could be detected with scientific instruments.
>
> Mephisto: Did you claim that gamma quanta have mass?
>
> AT: Established science states gamma quanta have no rest mass. That is easy to say since there are no gamma quanta at rest; they all move with the speed of light. If they had a rest mass, they could not move with the speed of light. But do they have energy? Yes, by all means. If they have no mass, is the equation $E = mc^2$

not valid? Hard to believe. But these are semantic problems, not really related to what we are trying to accomplish in this appendix.

The Point This Appendix Tries to Make

The *Region of Reality* is filled with *energy*, which is a substance, the only substance there is. Anything in this region is endowed with *mass*, which is linked to energy by being a property of energy. This means anything that does not have mass cannot be possibly found in the *Region of Reality*. The point we are trying to make is that one cannot expect to find the soul (if there is such) in the *Region of Reality.* The soul certainly has no mass. If it had such, it would have been detected by now by scientific sensors.

Appendix 2

The Dimensions

A. General Remarks

Why should we be bothered with dimensions in a book dealing with life after death?

It is indeed cumbersome to learn such an esoteric scientific subject matter, but there is a good reason for trying. In the sections describing the model of the world, we based everything on the assumption that there are different parts of the world, namely, *infinity*, *existence*, and *reality*. It was claimed that *infinity* has no structure at all, that *existence* has a vague structure, and that *reality* has a strict structure that is provided by the ordering schemes *space* and *time*. The fabrics of *space* are the dimensions. This is the subject of the present appendix. Still, why should we try to understand the details of the structure of *space* and *time*? The answer is that if science can make it plausible that there are these different parts of the world, then it is obvious that *reality* is "real." It would be also quite difficult to argue that the world part *Region of Existence* does not exist; obviously life is full of information.

Once science has made it plausible that there are parts of the world other than *reality*, it is also plausible that there may be more intangible parts of the world, like such as religion postulates. In section III on science and religion, we have seen that science cannot prove (or disprove) certain teachings of religion, but some of the teachings are plausible.

One cannot understand a concept unless one gets involved in it, so to speak, "kicks it around." Remember, one has to hear a subject matter at least three different times from three different sources before one can begin to *understand* that subject matter. For this reason, we recommend to try hard to understand the structure of *reality*, namely, *space* and *time*. Only then one can hope to grasp the more difficult concepts of intangible parts of the world like *existence* and *infinity*.

B. Ordering Schemes of Reality

Both space and time provide structure for the *real* world. Here, we want to make a most important point and claim that *space and time are ordering schemes*. Time orders the sequence of events that takes place (chronological order). Time, therefore, enforces the law of causality. The cause will always come before the effect. Space orders objects in relation to one another. An event can only happen if two objects exchange energy (see also appendix 1 on energy). In order for events to happen, objects have to come in contact with each other, whether this happens is determined by their spatial location. In case of *time*, cause and effect are ordered chronologically. In case of *space*, order in locations of objects ensures that a certain object A touches an object B so that an effect can happen. One should call this—analog to chronological order—"geological" order. But this word is, of course, reserved for a different meaning. Therefore, we will call it *topologica*l order, although this word is also taken for a different but similar meaning. Since the word is not as widely used as "geological," it will be acceptable to use it for our meaning as well.

As can be seen in appendix 1 (energy), there are three forms of energy that fill *reality*. One is *Potential Energy*, which extends in *space*, and the other one is *Kinetic Energy*, which extends in *time*. The third form, *Heat*, extends into both.

C. Space

Now that we sketched the function of *space* and *time*, let us go into some detail concerning the structure of space. We believe space has three dimensions. Maybe there is even a fourth dimension. If this is the case, the fourth dimension has to have characteristics similar to the other three. This means *time* cannot be the fourth dimension; it is a structure provider in its own right.

We rank the dimensions with order numbers starting with the "zeroth" dimension. (*Zeroth* is, of course, not a proper English word since order numbers start with first, not with zero.) This may seem a bit strange since we are all accustomed to the three Ds, namely, length, breadth, and width. But there is indeed a zeroth dimension, which is the "point." A point designates a location in space and, therefore, has neither length nor breadth nor width. A point on a piece of paper is, therefore, not a point but a blob since it definitely has a length, width, and height albeit all being very small. But this is a relative term. Under a microscope, it would look big smudge. A (mathematical) point representing the zeroth dimension would be still infinitely small even viewed with the strongest microscope since it has zero length, width, and height. Consequently, a point is invisible.

In summary, the zeroth dimension represents a point in *space*, whereby this point has zero length, width, and height. Such a point is also sometimes called a mathematical point. Its function is to designate a location in space. That means a precise location, not a range of locations, as a blob would be.

The *first dimension* is called a line, which is a structure that has only length but no height and width. Of course, a line on a piece of paper has also a width and, therefore, does not qualify as a line. A line representing the *first dimension* would be also invisible. A one-dimensional line, therefore, has to consist of a large number of points. These need to be "lined" up in a straight line like soldiers. Since the points have no length, one needs a large number of points to make up even a short line. To be precise, even a very large number of points would not form a line of any length. Rather, an *infinitely* large number of points is needed to form even a short line. It should be noted when lining up these (zero dimensional) points like soldiers, the condition to achieve the first dimension is that they all have to touch each other. (Above, we used the term "topological order.") Figure A5 indicates what we are trying to do, albeit with blobs rather than with points.

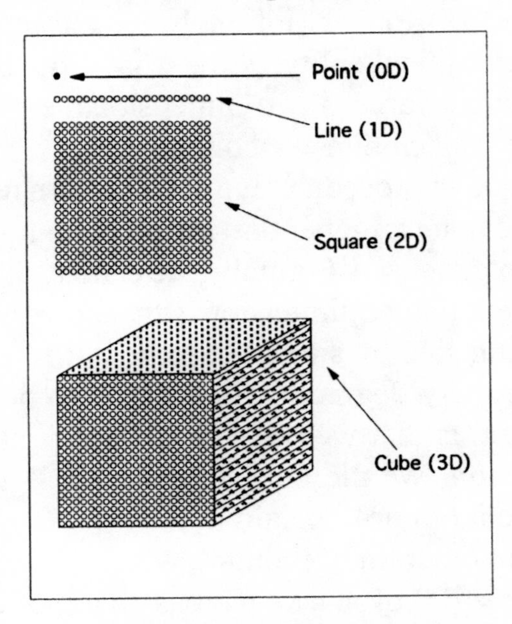

Figure A5: The Dimensions

How many points would then be needed to make a long line? Again, an infinitely large number of points. How about an infinitely long line? Again, an infinitely large number. This teaches us that *infinity* has no structure. It cannot be divided in substructures that can be counted. This is very important for understanding the *Region of Infinity.* We need to understand this in order to comprehend the message this book is trying to convey.

In order to obtain the second dimension, which is called an area, we have to invoke the same procedure. We take now an infinitely large number of lines (first dimension) and lay them side by side, like logs forming a raft (see also figure A5). Having done so without leaving gaps in between, we obtain the second dimension, namely, the area. Now we have an assembly of an infinitely large number of points in one row times an infinitely large number of rows, but the total number of points is still infinitely large. Note if we would have assembled this large number of lines not side by side but end on end, we would not have obtained an area but, again, a line, albeit longer (infinitely long) but nevertheless a line, a one-dimensional entity. Now, what did we do differently in order to obtain an area, a two-dimensional entity? We needed to align the lines side on instead of end on.

Doing the above procedure, we discovered a new dimension. What the new or a next higher dimension is cannot be described; it can only be experienced. Formally, we can say that the necessity to obtain the next higher dimension is that corresponding points in the lines lying side by side have to be under a certain angle. This angle is ninety degrees if a square is to be obtained. Yet what we just did is that we were describing how to obtain the second dimension. We did not explain what it actually is. We can experience it but not explain it.

To finish out the train of thought, we can now form the third dimension (the solid) by stacking areas on top of each other like one would stack pieces of paper (see also figure A5). Again, we need an infinitely large number of areas to form the three-dimensional entity, the solid. And again, if we would not have stacked the areas but laid them side by side, we would not have obtained the next higher dimension but just a larger area. Realizing all these, we now should be able to construct the next higher dimension—the fourth—by adding an infinitely large number of solids. But how do we add them up? Laying them side by side will not help. We will only get a larger solid. To understand why we are incapable to do this, we need to look again.

At figure A5, we see "points" lined up like soldiers. The next step shows lines laid side by side. The difference between the "lineup" and the "side by side" is a change in direction by ninety degrees (to obtain a cube). Now in the next step, we stack. The difference between "stacking" and "side by side" is again a change in direction by ninety degrees. In order to create the fourth dimension, we would have to change direction of addition again by ninety degrees. But we are out of degrees. Where are these ninety degrees?

	0 dim.	1 dim	2dim	3dim	4 dim?
0 D (point)	1	2	4	8	16
1 D (line)		1	4	12	32
2 D (square)			1	6	16
3 D (cube)				1	8

This means e.g the 2dim (which is a square) has ④ one dimensional entities namely points and ④ two dimensional entities namely lines. The four dimensional entity would have 8 cubes.

Now written as exponents:

0 D (point)	$1*2^{0}$	$1*2^{1}$	$1*2^{2}$	$1*2^{3}$	$1*2^{4}$
1 D (line)		$1*2^{0}$	$2*2^{1}$	$3*2^{2}$	$4*2^{3}$
2 D (square)			$1*2^{0}$	$3*2^{1}$	$4*2^{2}$
3 D (cube)				$1*2^{0}$	$4*2^{1}$

Figure A6: The Proportions of a Possible Fourth Dimension

Any ninety degrees that we could take would not produce anything new, just a larger solid. Does this mean there is no fourth dimension? Not necessarily. Maybe it is not necessary to change direction by ninety degrees for addition to obtain new elements (points, lines, and areas). But if this is so, we cannot imagine where this fourth dimension is because of our limited semantic levels.

In order to bring all these down to our semantic level, we introduce a trick, which we would like to call "skipping a dimension." Assume we invent a nation of peculiar people. What is peculiar with these people is that they are flat like bedbugs. This means they are two-dimensional; they have length and width but no thickness. We can get away with this because we assume tacitly that there are no significant changes in the direction of the third (skipped) dimension. Indeed, if we look on a person from the side, he resembles more or less a rectangle, with the exception of the face that is now seen in profile. Therefore, let us squash him flat that there is no thickness left. If we look at him from the front, he still looks very much the same as before. We have now a workable two-dimensional population. Now we can perform some *Gedankenexperiments* to get more familiar with such a situation. For example, let us put one of them in prison. For the 3-D case, we would put him in a cube that has steel walls (six walls: four sides, top and floor). The 2-D person we would put, therefore, in a square that has impenetrable walls (four sides). Can he get out of this prison? In principle, he should be able to do so by climbing over one of the four walls. However, he does not know that there is a third dimension, and therefore, he cannot find his way out. Does that mean the 3-D person can get out of his cube? He should be able to do so as well, if there is a fourth dimension and if he could figure out where it is.

This is an amazing conclusion, but it may strike most of us as nonsensical. Therefore, we will do now a real experiment, and we recommend highly that you actually do this experiment. If you do it, you will come to an amazing conclusion as well.

The experiment we want you to perform is called the Moebius ribbon. The reason we are doing it is to understand better how "skipping of a dimension" works. The skipping process removed one dimension from regular people to produce our flat people. We squashed them together so that their front and their back became one. They have now only one side since front and back are the same. To be truly two-dimensional, this is a requirement. The second dimension, the area, has four "sides," but these sides are lines, and lines have only a length but no breadth or width. Therefore, the area cannot have a top or bottom because it has no thickness (no height) since it is made of lines.

The Moebius ribbon has also only one side. To make such a ribbon is easy. Take a paper ribbon—e.g., from a printing calculator—and cut it to a length of about two feet. Now glue the two ends together with some sticky tape. But before you do this, rotate (twist) one end by 180 degrees in respect to the other. This will result in an endless tape, but of a somewhat unusual configuration, that we claim to have only one side. Since this is a rather incredible claim, proof is needed. One way to prove this is to try to paint one side of the ribbon but not the other. If you try to do this, you will find out that it is impossible since after the procedure is done, there is no unpainted side. Another proof would be to cut the ribbon in half—but lengthwise.

Before we do this, let's predict what should happen. If we take a regular sheet of paper and cut it lengthwise, we expect to obtain two separated halves. When we cut it, we know it is necessary to cut through two sides of the sheet, the top and the bottom. If this is difficult to see, imagine a plank of wood. Cutting the upper side of it (maybe an inch deep) will not separate the plank into two pieces, if the plank is two inches thick. Now we go back to our Moebius ribbon. We cut it lengthwise with a pair of scissors. If it had indeed only one side, it should not separate into two pieces after we are through with the cutting process. And indeed to your amazement, you will find out it is true. You will end up with one tape, not two, albeit longer and narrower than the original one but nevertheless one tape. One might say, if I cut this longer tape again, I must be cutting the second side, and I should obtain now two tapes. And indeed, if you do so, you will find again to your amazement that indeed two tapes are now obtained.

Having this now understood, we can go back to *Gedankenexperiments*. Now we take one of our flat people and amputate one arm, put him into the Moebius band, and make him go around once. If we amputated his right arm, we will find out—to our amazement—that after one round trip, he is missing his left arm, and the right one has miraculously grown back. The experiment is rather difficult to do, but there is a simple way to see why this must be so. Take a regular photographic slide and turn it around. You will see the same effect. If you own a slide projector, you have done this many times involuntary, and the effect is familiar to you. Did you ever consider how amazing this is?

Now let us attack the fourth dimension by "skipping one dimension." We take our population of flat people and distribute them over the surface of an inflated balloon. The surface of the balloon is two-dimensional (an area), albeit it is curved into the third dimension. However, our flat people do not know this. If they explore their universe by traveling in one direction, they will decide it is endless because they never can reach an end no matter how long they travel. They are correct in their conclusion, unless they also decide that their universe is infinitely large, which it is not. Their universe is curved into the next higher dimension (but not twisted like the Moebius ribbon).

Now let's fill their universe with galaxies. We accomplish this by painting a number of stars and galaxies on the surface of the balloon. The flat-people astronomers will discover these galaxies and will try to measure the distances from their own position to the galaxy under question. Next let us blow up this balloon slowly. The 2-D astronomers will now discover that all the galaxies are moving away from them. Not a single one will move toward them, and even stranger, the speed of flight is larger the larger the distance is from the astronomer's location. If the astronomers are smart enough, they will finally conclude that they are living in an expanding universe, whereby their known "space" (which is actually an area) is curved in the next higher dimension. Having concluded so, they can now compute the diameter of their universe, and since this diameter is increasing with constant speed, they can also compute when it all began (when the diameter was zero). Why do we tell you about such a crazy scheme? The answer is that this is exactly what our astronomers observe, and we could conclude we are living in a three-dimensional space, which is curved in a fourth dimension. The rate of expansion is called the Hubble constant. The time of the big bang could be determined this way. But bear in mind the center of the universe could not be found in the 3-D space. We could not travel to it with a spaceship because it is located in a fourth dimension.

Hubble discovered the red shift of spectrum lines in light coming from galaxies. Without going into detail, let's just state that a moving light source will display a red shift if it moves away from the observer, or it will display a blue shift, if it moves toward the observer. His discovery was that all galaxies show a red shift; consequently, they must be moving away from us. He also discovered that the farther away a galaxy is, the faster it moves. Originally, the theory above described was used to explain this. By now this theory is not the official theory anymore. Now the 3-D space expands in a way to conform to the observation.

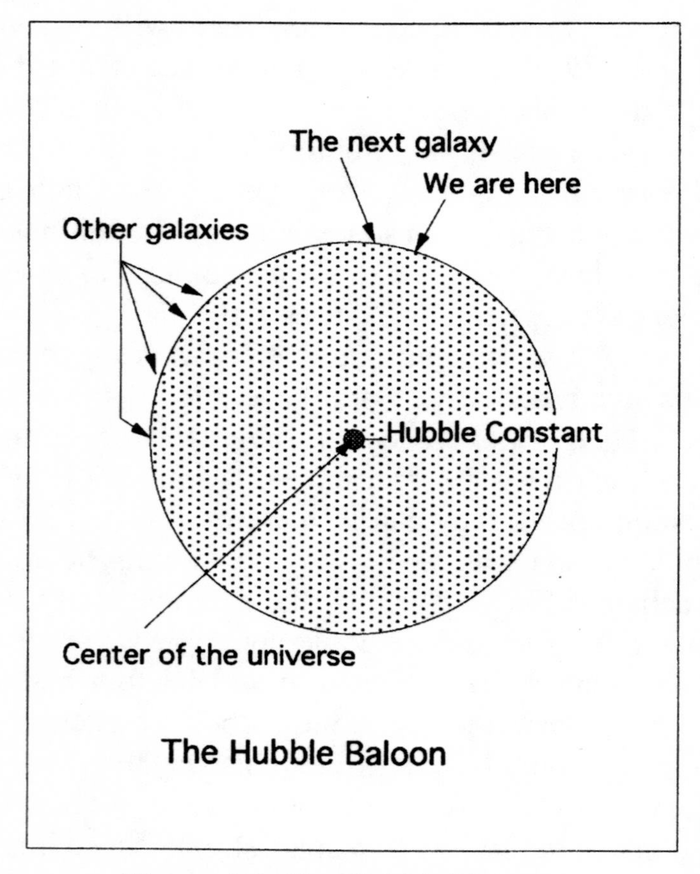

Figure A7: The Hubble Universe

This newer theory should not be discussed here since the connection to the subject of our book is not obvious. However, when discussing of the big bang (in another appendix), we will touch briefly on it again.

> Mephisto: You failed to mention that there might be other explanations for an observed red shift. One is to recognize the fact that the refractive index of space is not exactly 1.000. This would only be the case if there was no energy in this space. If so, space only exists but is not real. If it

contains energy, then it also contains mass ($E = mc^2$). Mass influences light. One of the first tests then of the theory of relativity was that light would be bent (attracted) by a mass. If light has to run against a gravitational field, it would lose energy (display a red shift). Don't you think there are plenty of occasions on the long way from the observed galaxy to Earth? Therefore, the red shift may be caused only partially by the movement of the galaxy.

AT: Nearby stars do move as measured not only by red shift but also by parallax. Ironically, this is called proper movement. It is not reasonable to claim that the galaxies sit still.

Mephisto: You should have said "sit still in respect to us." However, orthodox astronomy claims they move with a good fraction of the velocity of light away from us. Is that reasonable?

AT: Physics is not done by deciding what is reasonable or not but by what the measurements say.

Mephisto: The astronomer at the distant galaxy will claim we are moving away from him, and he sits still. He is also right since he observes a red shift imbedded in the light we are emitting. What is it now?

AT: The astronomers tell us that these galaxies are the oldest in the universe. It took the light so long to get here. To observe these galaxies, the astronomers call this the "look back" (in time). Again, we are being told that these galaxies were formed first after the big bang. We were formed much later.

Mephisto: Then tell me how we got here so far away ahead of the light. Did we move faster than the speed of light?

AT: That is impossible, and the talk of "look back" is silly.

Appendix 3

Theory of Relativity

A. General Remarks

Some say that only a few scientists in the world can understand the theory of relativity. AT does not claim to be one of them. Actually, the claim of difficulty stems from the general theory of relativity. Therefore, we consider here only a not-so-difficult subpart of the theory that is called the special theory of relativity (called STR in the following pages). This subpart of the theory of relativity is only valid for objects moving with a constant (not accelerated) velocity. "Not accelerated" also implies that the object moves in a straight line.

Why bother even with the easier-to-understand subpart? The reason is that the STR deals with *time.* Therefore, it is probably the most important theory that accepted science can offer for shedding light on the subject of the book. We want to know from this theory whether there is a beginning and an end of *time*. However, according to our stated policy, we will not try to teach you this subpart but only discuss the conclusion drawn by this subpart of the theory.

B. The Basic Premises

The STR teaches that there is no absolute space. In the ocean, there is a bottom onto which one can attach the anchor of a buoy. Using this buoy, one can tell where one is located in respect to the shore or other ships. One can even tell other ships how to find us by giving them our coordinates. These coordinates are in respect to the surface of Earth as an absolute location.

STR teaches that in space, there is no such absolute base one could refer to. Well, how about Earth? Assume we are in a spaceship one hundred million miles away from Earth. Indeed, we could tell everybody who wants to find us what our coordinates are in respect to Earth's coordinate system. However, both—Earth and our spaceship—move with constant (but different) velocities. Therefore, do we assume that Earth is at rest (it is not), or are we at rest? If we claim Earth is at rest, we have to assume that all other objects in space must be moving relative to Earth. That would mean Earth is the center of the universe. That is hardly a defendable opinion. Therefore, we may state that we move relative to Earth, but we have no way telling how fast we move relative to the empty space. There is no bottom where we could anchor a buoy. There is no scientific instrument that could measure our velocity relative to the empty space.

Before the theory of relativity, it was claimed that the empty space is filled with a substance called ether that does not move. This ether was supposedly the medium in which the electromagnetic wave called "light" moves. (Every wave is a vibration occurring in a medium: e.g., sound in air, ocean wave in water, etc.) Therefore, if we move with our spaceship and travel in the same direction of a light beam, the velocity of light we measure should be different from what we would measure when traveling against the direction of the light beam. It turns out that is not so. The velocity of light is the same in all directions. Ergo, there is not stationary ether. An experiment using Earth as a spaceship was done. It was arranged that at one setting, Earth moved with the direction of a light beam and, in another setting, against the light beam. This is the Michelson-Morley experiment. There was no difference detected between the two measurements. This rules out stationary ether.

There was also another experiment that produced totally unexpected results. In a forerunner of the TV tube, back then called a cathode-ray tube, it was discovered that the electrons in the electron beam (cathode ray) *seem* to have a different mass at different velocities. (Just to be accurate, it is actually the ratio e/m that changes with velocity. The symbol e means electrical charge of the electron; the symbol m means the mass of the electron. In principle, either e or m could change, but Einstein decided that it only is the mass that changes.)

Trying to explain these strange results, Einstein and some other contemporary scientists came up with some equations, which would describe what supposedly is happening. These equations are known as the Lorentz transformations and are the basis of STR.

C. The Basic Teachings of STR

* The STR deals only with linear (non-accelerated) movement.
* The world is a space-time continuum.
* All parameters describing space and time are relative; only the velocity of light is absolute.
* The laws of physics are the same in all inertial systems.

The term “inertial system” means an assembly of objects moving in the same direction at the same velocity. For the purposes of this book, we will be using a spaceship for explanation what such an *inertial system* would experience. Strictly speaking, such a spaceship should be far away from Earth, meaning not to be subject to gravity; otherwise, acceleration would take place.

The term "linear movement" means movement without acceleration in a straight line. (Flying a curve would require acceleration.)

The term "space-time continuum" implies that each point in space is defined by three-space dimensions (e.g., *x*, *y*, and *z*) and a time dimension (usually called *t*). A point defined in space without the time specified is meaningless. Consequently, each point has its own time. Simultaneity is impossible unless all four parameters are equal.

The laws of physics are valid for each inertial system. The transition from one inertial system to another one will not change the amount of energy. However, it might change the form the energy is in.

What does the word "relative" imply? For example, for *x*, *y*, *z*, *t*, and velocity *v*, there is no zero from which one could define the absolute value of these parameters. However, the parameter *c* (the velocity of light) is in all points of space identical, namely, 3×10^{10} cm/sec. (180,000 miles/sec.).

D. The Lorentz Transformations

The following equations specify the consequences of the above teachings:

$$t = \frac{t_0}{\sqrt{1-\left(\frac{v}{c}\right)^2}} \qquad m = \frac{m_0}{\sqrt{1-\left(\frac{v}{c}\right)^2}}$$

$$x = x_0\sqrt{1-\left(\frac{v}{c}\right)^2}$$

The meaning of the symbols used in these equations has to be considered very carefully. For a better understanding, and for the purpose of this book only, all coordinates in the above equations shall be understood as intervals. For example, t_0 is the time that passes between two events like two heartbeats. Once the inertial system moves at a different velocity, the time required for the same event to take place changes to time interval t, which is longer than t_0. This is usually referred to as time dilatation.

If the coordinate refers to a distance between two locations in space, namely, x_0 (but not y and z), it changes to x. For example, if there is in the spaceship a ruler that is parallel to the movement of the ship and is twelve inches long, it may now be eleven inches long. Of course, the marking will still read 12. This is usually referred to as Lorentz contraction.

Amazingly, this is also true for the mass m_0. Once the ship is moving with a different velocity, as it did before, its mass becomes larger. This is observed in the cathode ray experiment. The velocity v means the relative velocity between our spaceship and another one that moves parallel to us but with a different speed. The letter v designates the velocity relative to another inertial system (spaceship), but never could it be the designation relative to the space at rest since there is no rest.

If you are not familiar with square roots, just ignore them. For the argument we will make now, the square roots will have only a quantitative effect but not a qualitative one. Consider the first equation; here, the time interval t_0 is divided by a number $[1-(v/c)^2]$. Regardless of the size of v, the number in the bracket will always be smaller than one. Consequently, the time interval t will be longer than the time interval t_0. Also, the mass m will be larger than the mass m_0. Similarly, x will be shorter than x_0 as pointed out above (the Lorentz contraction).

In addition to the Lorentz transformations, there is also the most famous equation:

$$E = mc^2 \qquad \text{E: Energy}$$

This equation is Einstein's mass-energy equivalence. Among others, it states that all energy is endowed with mass.

If you are not familiar with algebraic equations of the above sort, do not worry. It is not necessary to understand how the above equations came about. One consequence of the Lorentz transformations can be easily seen. Namely, if something moves, length, mass, and time change. The "something" has to be another spaceship (*inertial system*). If there are no other such systems, we have no velocity that is relative to anything. Therefore, our ship will not become shorter and would also not become shorter if somebody were there to observe it. If there is another spaceship observing us, they will measure us as being shorter. Since there may be more than one spaceship observing us, plain logic demands that it cannot be our ship that gets shorter; it must be what the others observe. If they observe us, they will indeed measure that we are shorter, but these different ships will measure a different Lorentz contraction. The same is true for us when we observe the other ships. We also will measure that they suffer a Lorentz contraction. Of course, we will obtain a different result for each ship, depending on what their relative velocity (v) with us is.

E. Some Consequences

If an inertial system (spaceship) were to move with the velocity of light, v/c would be one, and the square root would become zero. Consequently, t_0 is now to be divided by zero. In this case, t would become infinitely large. Strictly speaking, the result of such a calculation (involving division by zero) is undetermined. However, if we were to make the velocity (v) just a tad smaller than the velocity of light (c), t still would become very large. Therefore, it is reasonable to assume that t would become infinitely large for the case that $v = c$. Therefore, a time difference between two events that takes one second at a certain velocity would take forever if the inertial system were to move at the velocity of light. In other words, time would not pass or at least pass very slowly. If v is reasonably small, this equation states that time will pass slower than time would pass in a spaceship at a lower velocity.

Considering the Lorentz transformation for the mass, one can see that the mass will get more massive (heavier) the larger its velocity (v) is. If it were possible to move it with the velocity of light, it would get infinitely large. This is the reason why the STR claims that it is not possible for any mass to move equal or faster than the velocity of light.

For the subject of the book, the Lorentz transformation for time (t) is the most important one. This equation actually states that the time passes slower in one spaceship as compared to the other that moves at a different (relative) velocity. Going to the extreme, if a spaceship would try to move with the velocity of light, its mass would get infinitely large, and time would not pass.

F. Difficulties

In the early 1920s, a journalist interviewed Einstein and wanted to know what happens when a body moves with a velocity close to or even equal to the velocity of light. Supposedly, Einstein told him that this body would get shorter and shorter each time its velocity is increased, and eventually, when moving with the velocity of light, it would disappear. (The logic is that when x is zero, it does not exist anymore.) We do not know if Einstein actually told the journalist this—only to get rid of him—or if he told him something else, and the journalist did not listen. In any case, it unfortunately got printed. Popular books and even some physics books repeat this nonsense to the very day.

In such publications, one can read that the inhabitants of the spaceship (inertial system) suffer a contraction in their x direction, but it is not measurable because the ruler used also shrinks. Provided the movement is in the x direction, only the x direction would shrink but not the y and the z direction. The confusion comes from the interpretation what x_0 means. It does not mean that this inertial system is at rest. The reason is that there is no rest. The theory of relativity teaches that there is no absolute space like ether that is in rest. Our inertial system may be in rest relative to another spaceship that travels parallel to us at the same velocity. But there are other spaceships that travel faster and slower than we. Which one counts for being at rest? We cannot use Earth since it travels as well.

Yet we could say forget all the other spaceships. Just compare what happens in our ship when it went from one velocity to another one. However, because we do not know what the velocity was before, how can we know what it is now? The velocity we would like to measure is the velocity of our ship in respect to empty space. The fact is that these velocities are not observable (measurable) because there is no rest. Modern thinking maintains that what is not observable does not exist. Consequently, only relative velocities to other spaceships exist.

Maybe we could make a comparison with the velocity of light. If we tried this, we would find out that the velocity of light is the same in all direction inside our ship, regardless of how fast we travel (really how fast others travel). The interpretation of the Michelson-Morley experiment suggests this.

Here is the main point for understanding all these: The Lorentz contraction (the shrinking of x_0) can only be observed from another ship due to its relative velocity difference with us. The same is true for the time difference and the mass difference.

To put it bluntly, if we look at the electron in the cathode ray tube, we observe it from our laboratory (*inertial*) system while the electron has its own *inertial system.* From our laboratory system, we observe that the electron's mass gets larger. However, if we were sitting on the electron, meaning we are in the same inertial system as the electron, we would not observe an increase in mass.

Most importantly, it should be pointed out here that inhabitants of any spaceship are not able to observe (or measure) a change in their parameters when the velocity of their ship is changed. The reason is that these parameters do not change. But what they can measure is their relative velocity with us, and that is the v to be inserted into the Lorentz transformations. Naturally, the acceleration between two constant velocities can be measured, but not the value of the constant velocity. Other ships that observe us can measure a relative velocity between them and us. This relative velocity is the v that needs to be inserted into the Lorentz transformations. Therefore, all other ships will see us to become shorter if our (relative) velocity is changed. The amount of observed contraction will be different for each ship, and this is regardless of whether they go faster or slower than we do. (The v is squared.)

The STR teaches that *reality* is a space-time continuum. This is the most important claim of the theory of relativity. In plain English, this means when dealing with an object situated in *reality*, the three space coordinates and the time coordinate have to be known. For each location in *reality*, these coordinates are different. Specifying a location only by x, y, and z is meaningless. One might say that time is the fourth dimension. However, that is no longer in vogue. What is important is the fact that, according to the STR, there is no simultaneity (equal time) for all points in space.

Well, this is a giant mess. No wonder that the journalist got confused. The reason for this confusion is that it is believed that the solutions of the Lorentz transformations are valid for the objects and inhabitants inside our ship. This is wrong; these solutions apply only for our observers. Consequently, we will not shrink to nothing because another spaceship that moves close to the velocity of light (relative to us) is observing us.

Now review the consequences of all these. A ship observing us will notice that we will become shorter when our velocity is changed. Inside our ship, we will not observe any change. If this other ship somehow can read our clock, it will notice that our time passes slower.

These seem to be far-out claims. Yet they are not so far out as it seems at a first impression. A common metaphor may show that things can take on a different shape when they are not really changing. Take a picture of yourself and look at it straight on. Now start to rotate it around an axis that runs from the middle of your forehead along your nose and through the middle of your lips. The more you rotate it, the more will your face become narrower until you cannot see it anymore. You know why this happens. You do know that you made a measurement, and nothing is wrong with this measurement. The object has, according to your measurement, indeed changed its shape. If you used a ruler taped to your picture, the ruler shrank too without changing the values printed on it.

A similar metaphor can be made for time as well. Consider a single lens-reflector photographic camera. These cameras have a focal plane shutter. That means a piece of cloth is pulled across the film. This cloth has a slit in it, the width of which can be changed. For extreme short-time exposures, a very narrow slit is pulled across the film. The consequence is that some parts of the resulting picture are taken at different times. If you were to take a picture of a fast-moving car, you would find out that the wheels are not circular but elliptical. Again, you understand what happens. But you made a measurement that is not wrong. You have obtained a demonstration as to what it means that each location in space has a different time.

Now let us take a side step and explain regular (nonrelativistic) relativity. This will help understand that observations of an event will appear different if observed from different observation platforms. Let's take a train having large windows so that an observer, standing on the ground outside the train, can observe an ongoing event inside the train. The event will consist of a tennis ball being dropped inside the train. The passengers inside the train will observe the ball falling down in a straight line. The "stationary" observer outside the train will see the ball falling down describing a parabola. This is quite a difference. If x is the length of the trace the ball describes, it is different for each case. If the event would be observed from still another train going at a different velocity, the trace that the ball takes is still different from the other two observations. This metaphor is very powerful, but it has shortcomings. In order to do it right one would need to do the experiment in a spaceship at zero gravity. In this case, the ball would not drop by itself but needs to be shot by compressed air out of a tube. After leaving the tube, its velocity would be constant. (In the train example, the fall of the ball is accelerated.) In the spaceship example, the ball would travel straight down with constant velocity on a straight path. The other spaceships would also observe straight lines of travel but under different angles, depending on the relative velocity between the ship that observes and the ship that has the ball.

So far, the velocities involved are relatively small. In order to observe "relativistic" effects, we would have to move with velocities that are a substantial fraction of the velocity of light.

Only in such a case, the time required to fall down would be different as observed by each inertial system. Also, mass of the ball would be judged to be different based on the observed traces although the original trace has not changed. From the point of view of the observing ship, the mass of the ball will change during the time required for the fall.

Mephisto: Are there other experiments that confirm this mumbo jumbo? And could it not be that Michelson-Morley is interpreted wrongly?

AT: Yes, there are others, but the determination of the truth is not based on the agreement of the result of most of the experiments. If the Michelson-Morley experiment is interpreted in the wrong way, then the theory of relativity needs to be abandoned. There is indeed underground literature that claims that, but this book cannot be a platform for discussing this. The majority of the physicists subscribe to the theory of relativity as being "correct."

Mephisto: The theory claims a body cannot move faster than the velocity of light, but there have been experiments performed where at least sixty different particles moved faster than the velocity of light.

AT: I know about this claim, but more work is needed before the scientific community will accept it.

Mephisto: Then explain the twin paradox to me.

AT: Since you bring it up, we do this in the next paragraph.

G. The Twin Paradox

Let's take an example. An astronaut leaves Earth in a spaceship and leaves his twin brother behind. To be correct, the brother left behind should also be in a spaceship, and both ships should be far away from Earth. Otherwise, the brother left on Earth could be considered as being at rest. Assume these ships move with constant but different velocities. Therefore, if the astronaut in the spaceship who was sent away indeed experiences time dilatation (his time supposedly passes slower), all events, not just the pendulum swing of a clock, would take place slower. This includes his heartbeats. His body counts these up, which is experienced as aging. Therefore, upon returning back to his brother who was left behind, he should be younger than his twin brother. This is called the twin paradox of the special theory of relativity.

Since the beginning of the formulation of the theory, exciting arguments have been going on whether this can be so. There have even experiments been performed to demonstrate that there is indeed such a phenomenon. Of course, these experiments were inconclusive. Elaborate mathematical systems have been developed to explain this problem away. Yet there is no such problem.

Actually, the astronaut could communicate with his twin brother by radio waves. He could send a message: "Today I turned thirty years of age." To his amazement, his twin brother would have to admit that—when receiving the message—he himself is actually thirty-four years old. Now is that really so amazing? All it means is that his brother, the astronaut, is now four light years away, and he was indeed thirty years old when he sent the message. Of course, it took four years for the radio wave to get to the brother who should factor this in when drawing conclusions from this message.

Therefore, concluding that the astronaut is younger than his twin brother left behind is not meaningful. First of all, it is not true that his time passes slower. It is the same as with the Lorentz contraction were x actually does not get smaller but is observed to be smaller by observers on a different inertial system.

Therefore, is the traveling brother actually younger? Such a comparison could only be done if both brothers could meet at the same location and time. This is not possible since the (v) in the Lorentz transformation is squared (minus v squared is the same as plus v squared; therefore, if he were indeed younger, then going in the opposite direction will not make him older). More plainly, the theory states that a location in the time-space continuum is defined by the 3-D coordinates and a local time coordinate. The brothers would need to move to a location where all the four coordinates (x, y, z, and t) are identical. It would take obviously one of the brothers (it does not matter which one) four years to do this. Having done this, they would discover that they are (by definition) of the same age.

Here Is a Preliminary Conclusion to All These

STR states that a specification of a location for an object is meaningless unless all four coordinates are specified. Therefore, there cannot be time without space and no space without time. They existed either always and existed always together, or if created, they were created together. Can they be destroyed? The theory of relativity takes no position on this.

We tried to explain the theory of relativity as faithfully and as comprehensively we could despite the many inconsistencies this theory is afflicted with. In the appendix on time, we show possibilities that are not necessarily taught by the theory of relativity but also do not disagree with it.

The question why we went through all these is "Did time have a beginning, and will it have an end?"

We obtained only a partial answer, namely, if time had a beginning, there also must be a beginning of space. If there was a big bang when energy was created, space and time must have been created too. Sure, they may have *existed* before the big bang, but they could have only had been *real* if they were filled with energy. (Remember, time and space are the hosts of energy.)

One might speculate that what is true for the beginning also must be true for the end. Such a speculation would not be proper since the initial condition before the big bang is certainly not close at all to the conditions when time would need to end.

Appendix 4

Einstein's Pool Table

Einstein used "skipping a dimension" when he tried to explain gravity. Although it seems there is no connection to the subject of the book, there is one after all. Einstein's pool table is a clever way to illustrate how gravity might be caused by a distortion of the fabric of space. The jargon used by the theory of relativity is the metric of space is changed by the presence of a mass. Changing the metric of space means that a straight line is no longer straight but becomes curved when a mass is put in the vicinity of this line.

The pool table metaphor goes as follows. Take a regular billiard table (one without holes) and remove the plywood underneath the green fabric. Make sure the fabric is taut. Then put a sphere of some reasonable weight (e.g., a grapefruit) in the center of the pool table surface. Naturally, the fabric is now distorted by the weight of the sphere. Now pretend this sphere represents the Sun. Therefore, take a marble (a planet) and put it anywhere on the pool table. The "planet" will immediately take off heading for the "Sun." Now give the marble a little kick in any direction except pointing to the "Sun." The planet will go in orbit around the Sun. If the planet is kicked in a direction parallel to the edge of the pool table, it should go into a circular orbit.

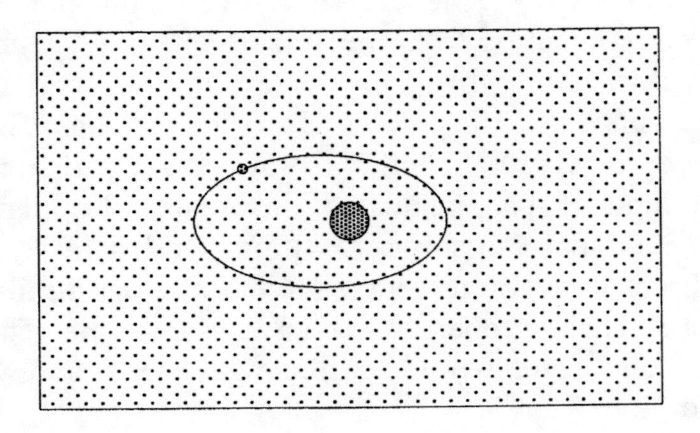

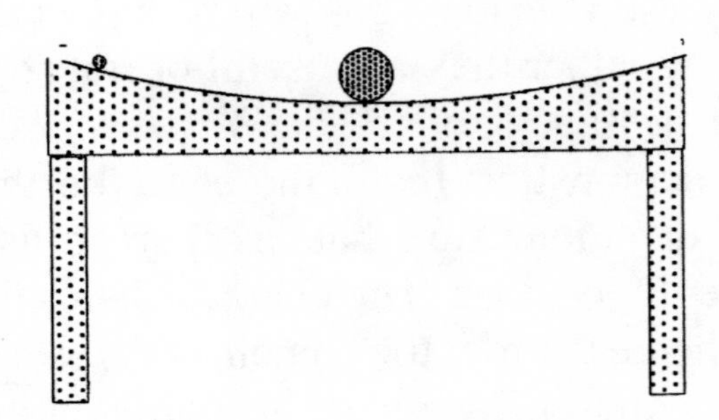

Figure A8: Einstein's Pool Table

Now why is this experiment an exercise in "skipping a dimension"? Strictly speaking, the sphere should be a disk, and the marbles should be disks too, but the friction between the planet disk and the fabric would not allow for it to go into orbit. But basically, the two-dimensional fabric represents the three-dimensional space. By the mere fact that a mass was placed onto the 2-D space, the (2-D) fabric of space was distorted (stressed). It means its metric was changed. If one drew a straight line onto the fabric before the "Sun" was placed there, it is now, after putting the Sun in place, a curved line.

The reason why the planet goes in orbit around the Sun is not that a force emanating from the Sun attracts it. (The grapefruit certainly does not do this.)

Like all analogies, this one also has shortcomings. The theory of relativity does not teach that the 3-D space is distorted into a fourth dimension as the analogy insinuates. It rather teaches that the 3-D space is stressed within itself. Clearly, such a fourth dimension would have to have properties similar to the 3-D space. These properties are energy and time. Therefore, the fourth dimension would be filled with energy and would be linked to time.

The point we are trying to make here is that such a fourth dimension would be *no* place where eternal souls could go.

The above-described ideas concerning higher dimensions are only theories. Whether they are useful or not so useful has no bearing on the message of our book. We described it here for the sole reason to demonstrate that if there is a fourth dimension, it is not "time" but a dimension that has characteristics similar to the other three dimensions. The characteristics of *space* (3-D) consist of a scheme to order the contents of *space* in topological order while by the entity *time*, the contents are ordered in chronological order. Again, this means that the entity "*time*" enforces the law of causality; it sees to it that the cause always comes before the effect.

Appendix 5

The Probabilistic Character of the Laws of Nature

A. Motivation

For any philosophy, including the one on life after death, the validity of the laws of nature is of utmost importance. In this appendix, we discuss a universally accepted position of science, namely, that the laws of nature are not valid in an absolute sense. Rather, what these laws predict to happen, will not always happen but only in a certain (albeit large) number of the cases. Therefore, the laws of nature give only a prediction of what *probably* will happen, but there is no *certainty* that it indeed will happen.

For anybody who assumes that the law of causality is the basis of all science, this will come as a shock. Therefore, one would expect that the scientific community considered all options very carefully before giving up on the law of causality. The law of causality states that every *effect* has a *cause*. The laws of nature prescribe the effect a particular cause will have. The effect so prescribed for a particular cause will happen every time (without exceptions). To put it bluntly, why have an exact science at all if future events are unpredictable by the laws of nature? Logically, an effect that happens without a cause needs to be categorized as a miracle.

However, to do science justice, the situation is not as black and white as insinuated here. It all goes back to the development initiated by Schroedinger's equation, which happened in the 1920s. This equation started a new development in physics that, and so to speak, rewrote the textbooks of science. To date, Schroedinger's equation has become such a powerful tool for computational physics that one could no longer do without it. The majority of theoretical physicists spend most of their time solving this equation for new applications.

New developments take about ten years to filter down to the teaching done at the high schools. AT remembers the 1940s when he was in high school, and the physics teacher talked about this new development, involving probabilities that the apple may fall one hundred million times to the ground, but at one time it will not—it might even rise. Actually, this is both right and wrong. What is right is that it will fall most of the events and will do something else only for an extremely small number of cases. However, even then, it will do certain things that are predictable by the laws of statistics rather than the laws of nature. What is wrong is the statement that the apple might even rise. While this may sound confusing, we will try to explain it in the following. For a physics teacher not doing state-of-the art research in theoretical physics, he got to the idea pretty close. However, to cause such an impression here would be unfair to what science actually teaches, although somebody with a sarcastic personality could actually put it this way and get away with it, as long as he avoids giving a specific number (like one hundred million). In order to be able to form your own opinion, which includes whether miracles actually can happen, we need to tell you the whole story in an understandable presentation. Therefore, please try to work through this appendix, and we promise you that after reading it three times, you will understand it.

What follows now is an explanation of a seemingly unimportant phenomenon, namely, *diffraction*, but we need to discuss it to be able to understand the consequences of Schroedinger's equation.

Diffraction describes what happens if a wave (of any nature) goes through an opening (or goes around an obstacle). It turns out that such a wave is distorted by such an opening in a rather mysterious way. Fine, but why is this so important? *Because it is the universally accepted opinion of contemporary science that all particles have also wave properties and all waves have also particle properties.* Again, this may be hard to swallow for the layperson, and we mention it here only to motivate you to work through this appendix. Obviously, since the world (*reality*) consists of nothing else but particles and waves, such a radical position has grave consequences on any aspect of our lives.

In order to help understand *diffraction*, start with figure A9. Here, a wave is shown that approaches an opening. This wave may be a surface wave on water trying to make it in through a narrow inlet. Therefore, a seawall with an opening is indicated in figure A9. This wave incoming from the ocean is considered to have wave fronts parallel to the beach. In figure A9, the seawall is also parallel to the beach. The thick lines in the figure mark the crest of the wave while the thin lines mark the valley (trough) of a wave. The direction of movement is perpendicular to the wave fronts.

You may already have heard that light consists of particles (photons). Be advised that *diffraction* is the classical "proof" for the wave nature of light. Unfortunately, there is also a "proof" that light is comprised of photons. This gives you an idea where the trouble comes from. In our discussion of diffraction, we stick with the wave nature of light and ignore its particle nature for the time being.

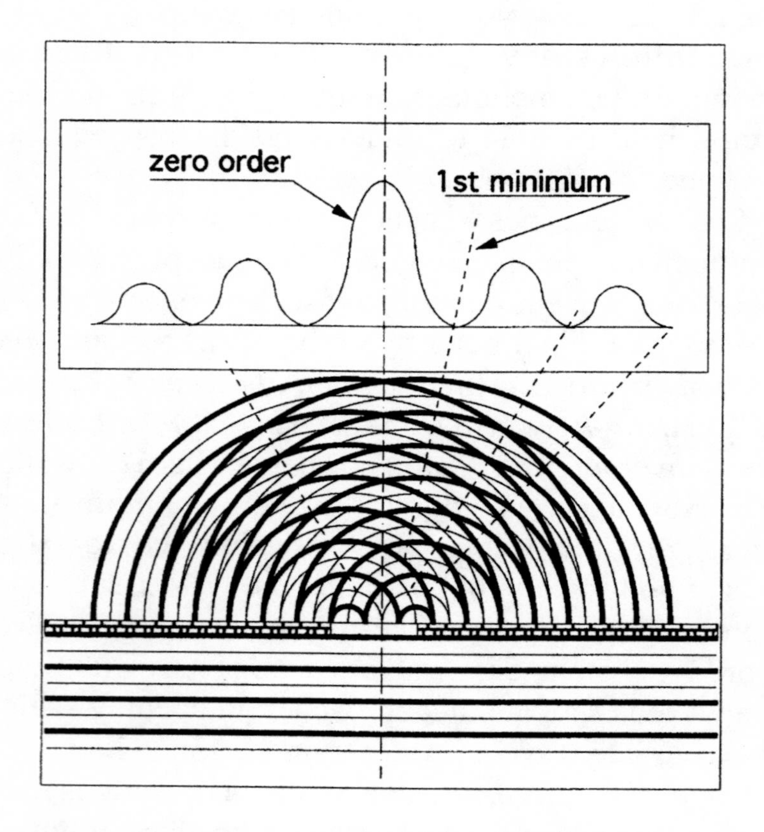

Figure A9: Constructive and Destructive nterference

For better understanding, we show in figure A9 surface waves of water, but there is actually no difference (as far as diffraction is concerned) from a light wave. For diffraction to show an appreciable effect, the size of the opening should be the order of the wavelength. (This is the distance from one crest to the next one.) By "on the order of," we mean the size of the inlet should be larger than one wavelength but smaller than one hundred wavelengths. For good measure in figure A9, we picked the inlet to be four times the wavelength. Figure A9 shows what will happen. The wave in front of the inlet continues through the opening while the rest is absorbed or reflected by the seawall. Assume there is an instrumented panel farther inside the opening. This instrumented panel measures the energy delivered by the impacting wave to it on various points. Then we plot, as shown in figure A9, the amount of energy versus the x direction. The naive expectation is that the observed shape should be a shadowgraph or the opening in the seawall. Surely, we do not expect sharp corners; therefore, we expect a lobe as shown as the solid line marked "0 order maximum." However, this is not the only place where energy is delivered. The plot at the instrumented wall also shows several side lobes (marked "side maximum"). The phenomenon causing such side lobes is called *diffraction.*

Of course, the next question is why. The answer is given by Huygens's principle. It states that the edges of the opening act as if they were emitting spherical (or in our 2-D case, circular) waves. Note the incoming wave was parallel (a jargon word that means the wave fronts are straight rather than concave or convex). Inside the opening, this wave seems to cease to exist and, in lieu of those, two circular waves that are emitted by each edge of the opening. This situation is depicted in the figure. Obviously, the two circular waves get in each other's way: they *interfere* with each other.

For *interference* to happen, two possibilities exist. In one case, the crest (amplitude) of one wave gets on top of the amplitude of the other. (Two thick lines intersect.) This is called *constructive interference.* In this case, the amplitudes of both waves are added together, and twice the energy is at this particular section of the wave. This has the effect that the amplitude of the wave is now twice as much as it was before *constructive interference*. In the other case, the crest of one wave ends up on top of the trough of the other. (A thick line and a thin one intersect in figure A9.)

Upon addition of the two amplitudes (one of the two is negative), the resulting amplitude is zero. In other words, the crest fills the trough with water. This is called *destructive interference.* At this particular spot, the water is now at the mean water level. This actually happens, and there is no point to argue with nature that this does not make sense. Of course, neither does Huygens's principle make sense. We say this because the two edges of the seawall are entirely incapable of emitting two powerful waves. They would have to shake to do this. If so, it should be dependent on the material of the wall how pronounced the effect is. It does not. This is even stranger when the wave is a light wave. Two pieces of black paper will certainly not be able to emit light. Nevertheless, a slit made with black paper will show the same diffraction pattern as a slit made of metal. The only requirement is that whatever is put there to provide an opening has to be able to stop the incoming wave. The water "acts" as if the edges of the seawall would emit a circular wave. To make matters worse, all points between the two edges supposedly emit such waves as well. (These were left off the drawing to avoid clutter.) However, Huygens's principle is a very powerful concept, if one needs to compute the shape of the wave pattern. Therefore, it is widely used. However, it is a very poor concept to explain what actually happens. The calculations that are done based on Huygens's principle can be verified easily by experiment. This tells us that the fact that something can be calculated correctly does not mean that the concept, on which the calculation was based on, is correct. In case of astronomy, after the heliocentric solar system was commonly accepted, some astronomers still used the rotating levers of the geocentric (Ptolemaic) system. It was easier to use and produced correct results (but the concept was wrong).

The main reason why we brought all these up is to get you familiar with the concept of *diffraction.* Therefore, it is now appropriate to define *diffraction*. It is the phenomenon that light deviates from its straight path of travel because it traveled through an opening, for example, a slit or a hole. Note, the light that ends up in one of the side maxima certainly did not travel in a straight line to get there. Why *interference* happen was explained above. At certain locations, wave crests are wiped out by wave valleys.

Now let's assume *light* is not a wave but consists of small particles. Will there still be side lobes on the instrumented panel? One would think not. Just to make sure, we will simulate this using Ping-Pong balls. They are shot through the same opening with a machine that shoots straight. Experiments like this have been done. Some of the balls may hit the side of the opening and are indeed deflected form a straight path. But this is random and not consistent enough to form side lobes. Therefore, one has to conclude that light is a wave and there are no photons. However, there are other experiments that (supposedly) only can be explained by assuming that light consist of photons, meaning it is not a wave. Orthodox science gave this situation a name: *dualism of wave and particle.* Meaning it does depend on the experiment for the light to behave as a wave or an assembly of particles. This is a generally accepted theory.

B. The Big Surprise

Up to the turn of the nineteenth century to the twentieth century, science was convinced that light is a wave. *Diffraction* was very well understood. But then the photoelectric effect was discovered. This effect, so established science concluded, can only be explained with the assumption that light consists of particles. Also, the fact that light is absorbed (and supposedly also emitted) in small energy packages ($E = h\nu$) points this way.

However, nobody could make diffraction go away. Therefore, the notion of a "duality" of the nature of light was proposed. A logical thinking student would assume that something cannot be a particle and a wave at the same time. It took several generations of students before this notion could be overcome. And nobody has yet cried out, "But the emperor has no clothes on."

To make things worse, at about the same time, people began experimenting with electrons. These are indeed particles (as long as they are free, meaning not inside the nucleus or atom). However, soon somebody discovered that when electrons are sent through a slit, a diffraction pattern is also obtained. (In the literature, usually a double slit, not a single slit, experiment is discussed, but there should be in principle no difference.) Supposedly, that could only mean that electrons can interfere with each other via *diffraction*—a wave phenomenon. (Here we go.) As discussed above, Ping-Pong balls certainly cannot do this. This could only mean that electrons also suffer from a "duality" of wave and particle. The conclusion must be that each electron drags a wave along, which tells it where to go, when going through an opening. The word "Führungswelle" (guiding wave) was initially used by Schroedinger for the wave that does this.

C. DeBroglie's Wavelength

If one has waves, one must be able to obtain images with appropriate lenses. So the electron microscope was invented. Of course, if there is a wave, there must be a wavelength.

This is called the DeBroglie wavelength and can be computed as:

$$\lambda = \frac{h}{mv}$$

(h: Planck's constant, m: mass of the electron, v: its velocity)

Where there is a wavelength, there must be a wave equation. This is an equation that describes the location of the energy (mass) carried by a wave in space and time. The wave equation for the DeBroglie wavelength was formulated by Schroedinger and is now called after him—the Schroedinger's equation. It allows computing the interference pattern for an electron beam that is experimentally obtained when going through a slit. Therefore, it is a very useful equation, and it is used in many different disciplines of physics and chemistry. In the opinion of AT, it falls in the same category as Huygens's principle, namely, a powerful computational tool but not subject to a physical interpretation.

Unfortunately, interpretation of Schroedinger's equation is exactly what contemporary science is trying to accomplish. According to Born's interpretation of Schroedinger's equation, the Führungswelle is not *real*; it only *exists* (in our terminology). For any wave, the square of the amplitude is the energy this wave carries. The square of the amplitude of the Führungswelle is the *probability* to find a particle at a certain location. This is Born's interpretation of Schroedinger's equation.

This is all fine and good, but why are we interested in this? The reason for this goes back to the late 1940s. The people who developed the earlier atom theory congregated around Niels Bohr, in Copenhagen. What they cooked up is now the so-called Copenhagen School. Their reasoning goes as follows: if an electron flies through such a slit, it is not clearly predetermined where it will go. Certainly, most of them will end up in this "0" order maximum (as in figure A9), but some will end up in one of the side maxima. There is no predicting which particular electron will go where. Only a *probability* can be given for where a certain electron will go. In contrast to this, the exact laws of nature require the electron to travel in a straight line, like a bullet, and not to fly around the corner of the slit. Ergo, all laws of nature have a *probabilistic* character. The law of causality is hereby repealed. Why did they do this?

To understand this, a few historical facts need to be known. At the end of the nineteenth century, it appeared that science could explain everything. The philosophical school of materialism seemed to win out. In this environment, a number of Christian intellectuals turned into atheists. Among those was Niels Bohr's father. However, his son became a religious fanatic. He felt he had to help out religion. Another member of the Copenhagen School was Heisenberg. He had also a bad conscience because he was the former head of Hitler's nuclear research. What was the issue?

The answer is the difference between determinism and indeterminism. The reasoning goes like this. If the laws of nature are exact, then everything that will happen is predetermined (determinism). This is so because everything depends on everything else. If just a pebble on the beach a million years ago had been at a different location as it indeed was, the whole history of the world could have been different. But now we discovered that the laws of nature are just a little bit inexact since some electrons can go where they please; no law of nature can tell them what to do. What they actually do can, therefore, not possibly be predetermined. Therefore, the whole history of the world cannot possibly be predetermined; it is actually determined as it happens. Therefore, any moral decision a person makes (here comes religion) is not predetermined by the architecture of the world but is rather made by this very person. Admittedly, this reasoning may have some holes, but that is the way it goes. That is what they mean by the probabilistic character of the laws of nature. As pointed out above, Newton's apple will still fall to the ground each and every time. Since the constant h in DeBroglie's wavelength is a very small number, the wavelength for a macroscopic body is unimaginably tiny. It can be disregarded for anything bigger than an atom. Consequently, in the case of Ping-Pong balls being shot through an opening, the side lobes of the diffraction distribution—albeit they are there—are so close to the zero maximum that the ball will deviate only an infinitesimal amount from the (straight) path that the laws of nature prescribe.

Mephisto: Are there actually such Schroedinger waves? Is there a physical proof for it?

AT: No, there is no physical proof. I learned quantum mechanics from an associate of Schroedinger's. He said that these waves exist only in the brains of the physicists; it is only funny that they still can interfere with each other in the real world.

Mephisto: Then why claim there are Schroedinger waves?

AT: Schroedinger's equation is so useful that if anybody came along and proved that there are no such waves, the physicists would probably shoot him with a shotgun.

Mephisto: What would Schroedinger say to all these?

AT: I never met Schroedinger himself, but I met his closest associate. Unfortunately, I was too young at this time to ask famous persons embarrassing questions, but basically the initial idea was to get rid of particles altogether. A particle like an electron would then be a wave packet. This is Schroedinger's interpretation of his equation. Unfortunately, Born ruined it by coming up with his interpretation that the amplitudes of the waves are the probability finding the particle at the location of the waves. For the average physicist, this is more exciting than a wave packet, and so the scientific community accepted the Born interpretation.

Mephisto: What is your opinion?

AT: Such probability exists (in our terminology) since it can be described by Schroedinger's equation. However, entities in existence are not endowed with energy. Therefore, they cannot cause electrons to deviate from a path they should take according to the law of causality. Consequently, the solutions of Schroedinger's equation describe something (correctly) but do not cause it. This is similar to Huygens's principle. Then what causes it? There have been proposals made by people in underground journals that the electric field carried by the electron has a structure that causes this behavior.

D. Let Us Summarize

If the laws of nature were absolute, everything that is going to happen would be predetermined, and there would be no free will to make moral decisions.

This problem is remedied by assuming:

1. the law of causality is invalid and
2. the laws of nature have only approximate validity.

Now that we know the story, we need to form an opinion on it. The consequences of the claims made are certainly staggering.

For us to ask is whether there are alternative explanations for what is observed. Start with *probability waves*. Do they exist? In our terminology, the probability waves *exist* but are not *real*. If we subscribe to the "Born interpretation," which is the accepted one, they cannot be real (probabilities have no mass). Yet they would be *real*—if one would follow the "Schroedinger interpretation." Remember, "real" means endowed with energy. Schroedinger's interpretation states that the square of the amplitude is actually the electron (its mass or energy) itself. However, if these waves, as claimed, are just *probability* distributions (Born's interpretation), then they are not endowed with energy (or mass) and, therefore, are not *real* but only *exist*. They are describable (with Schroedinger's equation) but, of course, not detectable. Since in the "Schroedinger's interpretation" the electron and the wave are identical, the wave would be indeed *real* and detectable, as detectable as an electron. Of course, what is detected as electrons may as well be a wave. To believe that electrons are little balls is naive. Therefore, how does one determine if something is a wave or a bunch of particles that have an inconsistent diameter? Well, one can do an interference experiment. Here we go—they show interference.

Whether the electron beam in your TV tube is comprised of particles or is a wave phenomenon cannot be determined experimentally if dualism of wave and particle is accepted. Then what is the reason why the scientific establishment chose to accept Born's interpretation rather than Schroedinger's interpretation? The answer is that with *probability,* determinism is no longer a viable philosophy.

Nevertheless, Born's interpretation modified the law of causality. The consequence of this is being felt in any scientific description of the *real* world.

Appendix 6

Entropy

A. First Approach

Entropy is a mysterious quantity very poorly understood by the average student of the sciences. If one asks the average physicist what entropy is, one gets an answer to the effect that entropy is a measure for the "order in the universe." The higher the order, the smaller the entropy. However, "order" is not observable; only matter/energy is. Therefore, there can be no instrument that "measures" entropy. If it is such a difficult concept, then why bother with it here? The reason is that in the following discussions, *entropy* will reveal itself as the main arbiter that decides whether or not things will happen. For this reason, it is important to understand the role of entropy as seen by official science.

Sometimes it is easier to describe what a certain entity is by saying what it is not. In this case, it should be pointed out that entropy is *not* energy density. One could surmise that the denser the energy is packed, the higher the "order in the universe." There seems to be some logic to this. However, let it be said here that energy density (the energy contained in a unit volume) and entropy are two different things.

Since entropy is not observable, it cannot be part of our physical world. And yet entropy determines whether or not events (e.g., chemical reactions) will happen. From this, one can see that entropy has to be a bridge between our physical world and some other world that is not physical but deals with possibilities (not probabilities) for events to happen, and therefore, one might even say ultimately it deals with fate. This means it is, therefore, a part of *existence*. Since entropy seems to be part of both worlds, *existence* and *reality*, let us try to understand entropy first in *reality*. Before we deal with the definition of entropy, let us look at an example that shows entropy in action. Assume that a tennis ball is being dropped from a certain height. It will fall to the ground and bounce back up to almost the same height from where it was dropped. To describe this event in terms of energy, first, the ball has *Potential Energy*, according to the height where it is. After being released, the *Potential Energy* is converted into *Kinetic Energy*—it moves. When hitting the ground, the Kinetic Energy is converted into deformation energy (which is a form of *Potential Energy*). The ball is deformed, but since it is elastic, this deformation energy is consequently converted back into *Kinetic Energy*, which makes the ball go back up until all its *Kinetic Energy* is converted back into Potential Energy, as indicated by the height obtained. The *law of conservation of energy* would require that the ball returns to its original height from which it was released (assuming it was released, not thrown). It cannot go any higher, regardless of how good or how advanced the material of the ball might be. In reality, it will not go quite as high as expected because at each of the described energy conversion processes, a small amount of heat is also generated. If the heat is summed up, the amount of *Heat Energy* will be exactly the same as the amount of missing

Potential Energy (indicated by the reduced height). After we understood this, we fill the tennis ball with water and drop it again. This time the ball does not bounce at all. What happens is that the Potential Energy is again converted into Kinetic Energy (it fell), but after hitting the ground, all the Kinetic Energy was converted to Heat. The water in the ball actually became warmer. This has to be so; the *law of conservation of energy* demands it. However, the *law of conservation of energy* would not be violated if the ball, lying on the floor, suddenly became colder and would rise back to its original height. However, we know this will not happen; what prevents it from happening is entropy. In other words, entropy determines that an event will not happen even though if it would happen it would not violate the *law of conservation of energy.*

For better understanding, we try another example. Take a coffee cup full of boiling water and pour it into a bathtub full of cold water. The bathtub water becomes only a few degrees warmer. However, there was no energy lost; it is all still there. Yet it is impossible to remove a cup of boiling water from the bathtub water although the energy is still there. This example shows that some energy conversion processes are irreversible. Entropy can be used as a measure of the degree of irreversibility.

Now let us try to understand the definition of entropy. As pointed out above, entropy is a bridge between the physical world (*reality*) and *existence*. Consequently, there are two different definitions for it. One in terms of *reality*, and one in terms of probability, which is chance and possibly, therefore, also fate. Let us consider the physical definition first. In this definition, the concept of temperature is involved.

Temperature is in itself a strange concept, although we are somewhat comfortable with it since we can "feel" temperature. We feel one object to be hotter than the other one. Intuitively, one would expect that the hotter object contains more heat than the colder one. However, this is not necessarily so. There is a fundamental difference between the amount of heat (which is a form of energy measured in joules or kWh) and the temperature (which is a relative measure of a property of heat). If one compares a coffee cup of boiling water with a bathtub full of body-temperature water, it turns out that there is more Heat Energy in the bathtub than in the coffee cup although the temperature of the latter is higher. The reason, of course, is that the mass of the water in the bathtub water is many times the mass of the water in the coffee cup. In case of water, the heat content is proportional to mass times temperature. Since *energy* is a *substance* and Heat is a form of energy, Heat is additive. This means if two amounts of water are mixed together, the combined (added) amount of energy is now present in the mixture, which forces the establishment of a common temperature of the mixture. This new temperature is determined by the combined new mass. Therefore, this temperature of the combined mass will be considerably less than the temperature of the coffee cup of water was but will be a little higher than the original bathtub water was. In other words, the temperature is not additive. The temperature scales were established before the concept of energy was discovered. The scale has been kept in the original form although it really does not fit in the physical measuring system logically. Basically, temperature is closely related to energy density. The concept of entropy is not related to energy density, although it is sometimes mistaken for it. Let us try now to understand entropy based on the physical (*reality*) definition.

B. Carnot Efficiency

At the time when steam engines were introduced, an engineer, Sadi Carnot, was trying to understand how *Heat* Energy is converted into mechanical (*Kinetic*) energy. The steam for his steam engine was produced by heating water with a coal fire. Therefore, the steam engine is a device that converts *Heat* into *Kinetic Energy*. He found out, to his dismay, that only a small part of the heat in the steam was indeed converted into *Kinetic Energy*. Therefore, he was trying to find out what fraction of the *Heat* Energy the engine would be able to convert into mechanical (*Kinetic*) energy and how much the engine would reject as waste heat. He soon discovered two important facts, namely, first, it takes a temperature difference for heat to flow and, therefore, for a heat engine to work. And second, a given amount of heat is more valuable at a higher temperature than at a lower temperature even though the same amounts of heat are compared (e.g., 1,000 joules). These 1,000 joules seemed to do more work at 500^{0}F than at 100^{0}F. By "doing work," we mean "can be converted into mechanical (*Kinetic*) energy." Assuming there are 1,000 joules of heat present, and if all this heat were converted into mechanical (Kinetic) energy, one would obtain 1,000 joules of mechanical (Kinetic) energy regardless at what temperature the steam was before the conversion. But the strange thing, he found out, is that it is theoretically impossible to convert all the 1,000 joules of heat into mechanical (Kinetic) energy; only a fraction can be converted. The heat not converted is not lost; it just stays heat. The fraction of Heat Energy that can be converted into mechanical (Kinetic) energy depends on two things: first, the temperature of the heat source and, second, the temperature difference between the heat source and the part of the heat, which is left over unconverted. This means that the steam that drives the steam engine has a temperature, that is, the temperature of the heat source; let's call it the upper temperature T_u; and the condenser, which condenses the steam back to water, has another (lower) temperature; let's call it the lower temperature T_l. What Sadi Carnot realized was that it is necessary to cool the condenser constantly; otherwise, its temperature would rise until it was equal to the steam temperature, and the steam engine would stop operating. In other words, there is a temperature difference (T_u-$T_{l'}$) necessary for the engine to operate. The fraction (*C*) of the amount of heat that can be converted into mechanical (Kinetic) energy depends on this temperature difference. Carnot found this fraction to be

$$C = (T_u - T_{l'}) / T_u.$$

C is called Carnot efficiency.

Nowadays one of the big bones of contention, when building a power plant, is that only a fraction of the applied heat (*f*) is used, and the rest is rejected and ends up in the river. This is the heat removed by cooling the condenser using the water of the river. This cooling process is necessary to maintain the temperature difference in the above equation. Typical numbers would be $T_u = 500^0C$ and $T_{l'} = 100^0C$. Converted into the absolute scale (*K*), the numbers are about $T_u = 800^0K$ and $T_l = 400^0K$. Therefore, the fraction *C* would be (800-400)/800= 0.5 or 50%, which means the "Carnot efficiency" is 50%. In other words, 50% of the heat can be (theoretically) converted into mechanical (*Kinetic*) energy, and the rest stays heat at the lower temperature. However, this is an ideal that cannot entirely be achieved. Modern power plants have about 40% efficiency. This means for each 40 kWh of electricity produced, 60 kWh of heat are dumped into the river. Consequently, if there were originally 100 kWh of heat in the steam at 800⁰K, only 40 kWh were converted into electricity, and 60 kWh ended up at low temperature (300⁰K) heat (to be cooled away in the condenser). Therefore, the remaining low temperature heat seems to be of lower quality as the high temperature heat is, although the quantity of the low temperature heat is larger than the one of the high temperature heat. This "quality" of the heat was defined as "entropy." There are ways to compute this entropy, but there is no way to measure it. Now the most important fact:

Energy conversion events from one form into some other form of energy will only happen if the entropy is increased in the conversion. Crudely speaking, the increase in entropy is defined as the amount of heat converted divided by its temperature. Since entropy is not measurable, there is no absolute amount of entropy defined. It is always increase or decrease of entropy that counts.

In summary, entropy is a mysterious entity that has to increase if a reaction or event ought to take place (by itself). How do the reaction partners know or how does the steam engine know whether or not the entropy is increasing? The same question could be asked for gravity. How does the apple know in which direction it is supposed to fall? But here you might say (because you were taught so) science has a splendid answer. Namely, there are gravitational field lines that are vector quantities. This means they point in a certain direction (the center of Earth), and the apple follows these field lines. However, there are "really" no field lines; they are just figments of human imagination. This in turn means that what is described by the field line formalism is just a more sophisticated way to describe the fact that the apple falls to the ground. This is a description, not an explanation. Of course, there is this other principle, which states that all systems seek to assume the lowest possible amount of Potential Energy. Certainly, the apple will have less Potential Energy on the ground than it will have on the tree. But why does a system want to assume the lowest amount possible? The answer—strangely enough—is entropy.

(Note what follows now cannot be found in textbooks of thermodynamics. However, our reasoning is straightforward, and you can judge for yourself whether it is valid or not.)

C. Entropy as an Arbiter

If allowed, a system will convert part of its Potential Energy into Kinetic Energy by itself. The fact that it happens "by itself" (meaning no energy has to be added to make the event go) indicates that an increase in entropy will result due to the energy conversion process. Now why? Because if an object moves, it sooner or later has to collide with some other object (in this case the ground). A collision means generation of heat, which means also the establishment of a temperature for this heat that in turn is reduced (cooled away) "by itself" either by conduction, convection, or radiation, which means an increase in entropy. (In every one of the three cases, the energy is distributed all over the universe, meaning the order of the universe is reduced or the entropy is increased.)

> Mephisto: Let me interrupt right here. How can you say that spreading out the energy all over the universe decreases the order in the universe (or increases the entropy)?
>
> AT: Look at a pool table. You can line the balls up neatly into a triangle. Each ball touches two other balls. But when you hit them with another ball, they will spread out all over the table in a more or less random fashion. Don't you think that the order in the distribution of the balls is reduced?

Now going back to the textbooks of thermodynamics, we find the definition for entropy in physical terms to be $dS = dQ/T$. Here *dS* means change in entropy (decrease or increase); *dQ* is a difference between the amount of heat, inputted and rejected, and *T* means temperature at which the energy conversion occurs.

For those who are not mathematically inclined, do not panic. This equation just says that the increase in entropy is equal to the amount of heat converted divided by the temperature at which the conversion takes place. We have to admit that this explanation in English is not any more enlightening than the mathematical equation. But remember, entropy is a strange quantity. Therefore, hang on; sooner or later, it will become somewhat clearer. Remember, a physical law, as described by the above equation, is nothing else but an instruction how to measure the quantity defined. An equation does not explain what it is and why it is so. Therefore, to put it in the simplest terms, any time we withdraw heat from a system and try to convert it (e.g., into Kinetic Energy), we find out that not all the heat is convertible. Some of it must stay heat (albeit at a lower temperature).

The alternate way to describe entropy, namely, that it is a measure of the order in the universe is really not in conflict with this. Since *entropy* can only increase for an event that happens by itself, the *order* can only decrease at such an event. Such a statement infers that there is a certain predetermined outcome of events. This is akin to fate. However, what is not true is that the order always will decrease at each and every event (despite the fact that entropy always will increase). Obviously, life evolved and did not start with a high state of development. Therefore, the identification of entropy with order is at least incomplete if not inappropriate. (Such identifications by a model are never wrong or right; they are useful or useless.)

D. The Statistical Definition

We describe now the other definition of entropy, the statistical definition, which is accepted by official science as well. Let us assume we take a transparent cylindrical container and fill it with white Ping-Pong balls. Then we put a layer of black Ping-Pong balls on top of it. Looking through the transparent walls, we can see white and black balls in clearly separated layers. After we rotate the container a few times, we will see that the black and white balls are now intimately mixed. However, if we start with a mixture of black and white balls and rotate them, they will not separate into two-layered portions of black and white balls. For this reason, if you ever would find an assembly of black and white balls left by somebody else in the container, it is more likely that they are mixed rather than separated. The same has to be true for smaller balls, even as small as atoms. Therefore, for a container filled with two different types of gases (e.g., hydrogen and oxygen), chances are that they are mixed and that they will not separate spontaneously into two separate regions. Therefore, there is a certain probability to find a system in a mixed condition, which is much greater than to find the system in a separated condition.

Let us say the container has the size of one cubic foot and filled with a one-to-one mixture of hydrogen and oxygen. This gas has a certain temperature and, therefore, a certain (amount) of heat. Now we insert a dividing wall to make it two containers. We remove the gas mixture and fill each of the now-separate containers with a different kind of gas, e.g., one with hydrogen and the other with oxygen. Each gas has a certain temperature and also a certain amount of heat. Now we remove the separating wall between the two containers so that it becomes one container again. We know that the gases will mix. This has to mean that the entropy of the mixed gases is larger than the sum of the entropies of the individual gases in a separated state. Probabilities are not additive; they rather are multiplicative. Entropy, however, is additive. If we want to define entropy in statistical terms, namely, by a probability, this description has to be compatible with the physical description, which is additive.

E. For the Mathematically Inclined Readers

$S = f(e) + f(y)$ is physical description. $S = p(x) * p(y)$ is statistical description. Since both equations must describe the same thing, we are looking, therefore, for a relation that produces a product (as in the statistical description) from an addition (as in the physical description). Such a relation is the logarithm. Namely, $\log(a * b) = \log(a) + \log(b)$. It is customary to use the natural logarithm. And therefore, the statistical description is

$$S = k * \ln(P).$$

Here, P is the probability of finding a "system" in a certain state, and k is a constant, known as Boltzmann's constant. S is the entropy.

Statistics does not deal with the laws of nature. On the contrary, statistics are used to predict that events will happen without consideration of the *law of causality*. The statistical definition of entropy does not contain temperature. Therefore, the "system" does not have to be a mass. It can be anything as long as it is comprised of a large number of individuals. So we crossed the line now. As long as entropy was restricted to systems comprising mass/energy, it applied to *reality*. If entropy now applies to assemblies of entities of any kind, this means it comprises information as well. Since information rests in *existence*, we have shown now that entropy applies to both *reality* and *existence*. This is the reason why entropy cannot be measured, and there is no absolute value for entropy (meaning entropy is not a substance).

As pointed out above, at any energy conversion, the entropy has to increase or (ideally) not change. In order for a system to experience a reduction of entropy, energy has to be added to this system.

F. A Different Look at Entropy

From the previous section, we know that the world needs a certain degree of nonrandomness in order to survive as a structured entity. Nonrandomness we usually call "order." Increase in entropy destroys order. In our terminology, increase in entropy removes already-*realized* information. For example, take a book—maybe as a typeset printing plate. Assume you remove one letter from its place and put it arbitrarily somewhere else. *Arbitrarily* means random. You have now increased the entropy of the book. This entropy content was very low to begin with, since a book is sheer information. But if we continue to do this, the book will become more and more a random assembly of letters until eventually no information is left at all. If we remember now that the spiritual world (*existence*) thrives on information and *real* world thrives on structure (order), we understand that the driving force of the spiritual world will have to make constant efforts to keep entropy from increasing. The very fact that the world exists (or maybe that we exist) proves that *existence* is successful in doing so. It may not prevail forever, but at least it does for the time being. Remember, in many mythologies, the death of the gods is a forgone fact (Götterdämmerung). They are subject to fate (entropy). What we need to understand now is how to connect fate and entropy. So far, entropy is just the ratio between Random Energy to Nonrandom Energy. We know already the connection to order. We will now show there is also a connection to probability.

In order to do this, we have to divert somewhat from the subject. For this reason, let us talk about traffic accidents. We can predict to a fair degree of accuracy how many people will die next weekend as traffic fatalities. Why can we do this? Because we believe that the occurrence of traffic accidents is random. Therefore, we say what has happened before under similar circumstances will happen again since what happened was random—it had no particular reason for it to happen. Therefore, if it is the same time of the year and if the weather is similar as it was at that time, we would expect the same number of people being killed. It sounds so reasonable that we almost believe it. What is claimed, however, is that the traffic accidents are events like other events, for example, a collision of two molecules. They will happen in the same random way. If all traffic accidents to happen next weekend would all happen at one place—e.g., in Niceville—and no accidents would happen anywhere else, that would be an extremely unlikely situation. Because the entropy of such a system would be extremely low. It is like saying all the molecules in a gas are in the same location, and if they are moving, they move all in the same direction. This means low entropy. It also means high structure or order. What entropy dictates is that they are all in different places and move all in different directions. Since we believe that traffic accidents are events like all other events and are governed by entropy, we predict that next weekend not all accidents will happen only in Niceville but they will be distributed randomly all over the USA. That is great, and it turns out to actually occur this way. However, what really is being said is that for a traffic accident to happen, randomness and probability are the determining factors, not the *law of causality.*

Let us explore this with an example. Assume a man gets up in the morning, eats his breakfast, and goes to his car. He starts it and drives to work. Halfway there, he crosses an intersection and collides with another car and is killed. If he had gotten out of bed only a minute later, he would have missed the other car and would have made it across the intersection unharmed. For him to die, it took a sequence of causes and effects according to the law of causality. Everything, which happened, had a reason. One could say that if there was anything at all that was random, it is the timing, namely, the time when he got up and started the chain of events. Now comes the strange thing. In principle, that would not preclude that all traffic accidents will happen in Niceville and nowhere else. What must have happened is that someone or something arranges the timing of all these causal events in a way so they do not lead to traffic accidents in Niceville only. In other words, if an accident has already happened in the morning of this day and if this were the quota allowed for Niceville (according to probability or randomness or maximum entropy), this somebody or something must time all other events in Niceville in a way that the people living there do not get in any further trouble. This is, of course, an incredible story because this way it is claimed that causal events (events dictated by the law of causality) happen at random. This is, of course, a contradiction. Yet the strange fact remains that the events somehow seem to know if an accident already happened in Niceville and prevent, therefore, that not more or at least not all accidents will happen in Niceville. Primitive people made human sacrifices. Does that mean they believed if they fulfilled the quota with a slave, the ruling class would be spared?

"Probability" is the credo of our times. With talks like this, one gets physicists, mathematicians, and statisticians extremely upset. They will probably burn the book. They just love to calculate probabilities, e.g., the probability that you will be hit by a meteor or that you will be killed by a nuclear power plant. They will tell you that these accidents happen randomly.

In contrast to this, we claim now that if you consider the traffic events as a system, the accidents will happen in a way that entropy is maximized. This means the events are distributed all over the USA like the gas atoms and their velocities are distributed in the entire volume of the gas. Even in the case of gas atoms, that does not mean that there is no causal reason for them to collide. They, of course, move according to the laws of nature, as is the case for movement of the participants of traffic. However, in both cases, the complexity due to the multitude of event is so great that we cannot trace all that is happening due to the law of causality. We rather cheat and invoke the laws of statistics. For practical purposes, this is fine. But for dealing with philosophical matters, this is, of course, nonsense. What we have to understand is how all these strings of causal events are synchronized to produce the correct quota of accidents. People who die in a traffic accident have met their fate. We have established now a connection between fate and entropy. The fact that our man got up exactly at the right time to meet the car that killed him at the intersection was determined by his fate. You may say that is not so; he got up because his alarm clock rang at a certain time, and he set the clock the previous night. This may very well be true. Therefore, we have to qualify our statement. There is, of course, a continuous chain of causal events since his life started. Therefore, what is timed by fate is not the beginning of a subchain of events but the synchronization of the total chain of all the events in his life with the chain of events of all other people or even inanimate objects. (One could be struck by lightning that has a causal reason why it strikes the ground.) The statisticians will say, "This is a large number of events. The interplay is random. Like if you shuffle a deck of cards, there will be always a card adjacent to another one. If the combination of king and queen adjacent to each other would mean a traffic accident, you will find this combination randomly distributed in the deck (especially if it is a stack of many conventional decks). If you would have a deck of a hundred thousand cards, you may find this combination may exist a hundred times, and you know for sure that these combinations will be distributed all over the deck and not all will lie next to each other. For the same reason, not all traffic accidents will happen in Niceville. This is entirely true, but it is no explanation. It merely states the facts. The fact that not all king-queen combinations are in sequences is as strange as is the fact that not all traffic accidents happen in Niceville. It still remains that somebody or something must see to it that a "random" distribution is established. Everybody is sure it will be this way, but there is no explanation of a casual nature why it is so. Our explanation is that entropy synchronizes all events in the world in a way that entropy assumes the maximum value that any combination of options could lead to. How is this accomplished? Is there a god who regulates this, or is this a mindless principle that exists and does its regulating according to chance? The answer will never be known. This is a matter of belief.

One could also take a step back and say that the transfer of information out of existence into reality is controlled (synchronized) by *existence* (see also the appendix on transfer mechanisms). Again, that can be a mindless mechanism that is just seeing to it that the correct number of people (the quota for Niceville) gets killed rather than to care about who gets killed. Or is it so that this synchronization done by *existence* is steered by divine intervention? It is a matter of belief.

> Mephisto: You spent a large number of words to describe a difficult-to- understand phenomenon that actually has nothing to do with the subject of this book.

> AT: The connection is not obvious. What was shown is that there can be a connection made between entropy and fate. Entropy is a part of physics, and fate is a part of religion. To my knowledge, this connection was never proposed before. We demonstrated so that there are indeed parts of the spiritual world that can be touched by science. If this is so, the statements on the subject of the book that are based on science become so more plausible.

Appendix 7

The Big Bang

The reason why we are interested in the big bang is that we desire to know if there was a beginning of *time*. It is the accepted opinion of official science that there was a big bang when the universe was created. This theory competes with other theories like continuous creation of the universe or the reordering of the chaos of an already-existing universe. We submit that it is reasonable to conclude that if there was a beginning of time, there might be also an end of time. This, of course, is in violation of the idea of an eternal soul.

In 1951, Pope Pius XII declared his approval of the big bang theory. Of course, he had advisors, and some of these were astronomers. While this decision may have been made on purely theological contemplations, this declaration has certainly a large influence on secular scientists.

In case the universe was created out of nothing, this would violate the *law of conservation of energy*. However, if the big bang was created by divine intervention, such an argument does not apply. Undoubtedly, the decree of the pope plays a very important role.

Therefore, let's conform to established opinion and accept that there was a big bang. What can be concluded from this? Since there was no energy before the big bang, there was certainly also no *Region of Reality*. *Time* is a part of this region. Consequently, there was also no *time* before the big bang. There might have been a *Region of Existence*, but we agreed already that the ordering scheme "time" is not a part of this region. The same is true for *space*. The conclusion is that there was no *time* and *space* before the big bang. Energy and mass were created in violation of the *law of conservation of energy and mass*. Consequently, the *Region of Reality* was created at the instant when the big bang happened. The word "bang" implies that this event happened fast. Since there was no *time*, it could not have taken any *time* to create the total amount of energy the universe of today is made of. Surely after the energy was *realized*, it could take time to form subatomic particle and then atoms and molecules, which in turn formed stars and galaxies. This process is going on even as of today, several billion years later.

There is as substantial number of people that do not "believe" there was a big bang. We put the word "believe" in quotation marks. The reason is that these people question the big bang theory on scientific grounds, and rightly so. A very well-known astronomer wrote a book on this subject showing the impossibility that a big bang could have happened. The scientific community ignored his book. Also rightly so. The point is that this is not a case of scientific theory but a matter of belief.

This book is not about scientific disagreement or about belief. Both certainly would go beyond the focus of the present book. But nevertheless, we want to know, and we are entitled to know, if there was a beginning of *time*. And if so, will there be an end of *time*? Why are we entitled to know this? Because we the people support both, science and religion, financially.

A suggestion could be that if there were an end of *time* and if souls exist, they would not be eternal. In the present book, we claim that the *feeling of I* is the image of the soul that rests in *existence.* We also claim that there is communication between the *feeling of I* and the contents of the *Region of Existence* via intuitions. We also claim that the *Region of Existence* does not contain time. The question arises whether there was a *Region of Existence* before the big bang. That could be, but it would be a mute situation if there were no *feeling of I* to communicate with. A reasonable suggestion is that if a creator who created the immense amount of energy as well as *space* and *time* created indeed everything, meaning also the *Region of Existence.* In other words, according to this reasonable suggestion, there would have been a beginning of time and a creation of souls when the big bang happened.

One might also speculate that the soul (or all souls) only *exists* as long as there are people alive. Nothing said in this appendix has a scientific basis (maybe with the exemption of the big bang theory) but is pure but reasonable speculation.

> Mephisto: How can you possibly claim there was no *Region of Existence* before the big bang? This region contains only Information, and no time and no energy. It could have been there forever.

> AT: You are right, but that is also true for the *Region of Reality*, and we would not need a big bang at all. Consider this: You buy a puppy when it is still very young and take it to your house. After it is grown up, the dog knows the house very well. Will it ask itself where this house came from? Probably not, since the house was always there (always as far as the dog is concerned). If it is a very smart dog, it might ask

itself this question, and it necessarily will come to the conclusion that a higher power produced it (humans). It would consider anything to be a higher power that can accomplish what a dog, or even a thousand dogs working together, could not possibly accomplish.

Mephisto: Fine, that is true for the house, but is it also true to for the information resting in the library?

AT: The dog does not know about this, but if it did, that would put it only at a higher semantic level, but that would again not be the end of it. Therefore, for the human case, we believe that a creator so powerful that he could create all the energy we find in the universe, in (literally) no time, certainly would not have to rely on a still-higher power to create the *Region of Existence*.

Appendix 8

Ghost, ESP, and Other Mysterious Things

What Mysterious Things?

Having gone into substantial detail on the science side to help you, the reader, to form an opinion on life after death, AT felt it would be a cop-out to avoid talking about ghosts and related things, although some of these subjects are not directly related to the subject of the book. Therefore, here is our opinion on ghosts, ESP, and other mysterious things. What are these things? Let us make a list.

ESP (extrasensory perception)
ghosts
spiritualism
clairvoyance
astrology
stigmatization
miracle healings

A. ESP (Extra sensory Perception)

ESP means to sense something without the use of sensing organs implying *realization* of information without exchange of energy (without *energy conversion*). Somehow information must get into the brain without energy exchanges. In normal sensory perception, sensors convert energy (light, sound, pressure, etc.) into nerve signals that are energy. Once inputted into the memory, the content of the information is contained in certain locations (addresses in computer jargon). The process to input information into the brain (involving sensory organs and, therefore, energy) we may call *sensations*. However, we need to admit that, in addition to *sensations*, there are also transfers, which we call *intuitions*. These are a direct transfer from the *Region of Existence* to the brain. The mechanism of transmission has been described in section VIII on transfer mechanisms. It is energy that is transferred, albeit in a fuzzy way (see uncertainty relation). This transfer is a short burst of a limited amount of energy carrying some information.

Let us take a typical example of ESP. AT has personal knowledge of this event. A woman, a mother of a teenager, woke up at two o'clock in the morning because she heard her teenage son calling for her. However, the son no longer lived at home but lived by himself in an apartment. After awakening, the mother realized this and believed it was a dream and went back to sleep, although with a bad feeling. The next day, a Sunday, she tried to call her son on the phone. Since there was no answer, she sent her husband to the apartment. Nobody was there. Now the bad feeling came back to both of them. They went to the marina to look for his sailboat. The boat was not there. Now they were really alarmed. They went to the coast guard station. The coast guard found the boat already washed ashore. They told the couple it took about two days before the body was washed up.

However, in this case, the body was found floating at the scene of the accident by another boat at the same day. Therefore, they could pretty much reconstruct what happened and when. It must have happened in the early morning hours of this Sunday, at about the same time when the woman heard her son calling for her. He obviously fell overboard, tried to swim, but not being a great swimmer, eventually drowned.

Stories like this are legion, too many to disregard all of them as coincidences. Therefore, how do we explain this? The information that the accident happened is certainly available in *existence*. Could it have gotten into the mother's brain out of *existence* by *intuition* transfer? If we agree that intuitions are possible, we have to agree that this type of ESP event is possible. Like *intuition*, they come out of nowhere and are short bursts of information. Like *intuitions*, also coming from nowhere (namely, *existence*), they are not bound to any location or time. But controlled experiments on this subject, e.g., reading cards, should not be possible since these are not short burst of information coming out of nowhere.

B. Ghosts

Traditionally, ghosts are transparent. They supposedly are apparitions of persons once alive. Such a person may have died hundred of years ago in a violent death and cannot find rest until justice is done. Albeit only a part of the population believes in ghosts, a state supreme court (NY) nevertheless recently found that a certain house is haunted. Therefore, according to the court, ghosts have to be taken seriously. Our position is, therefore, that ghosts *exist.* Note that the word "exist" is printed in italics, having therefore a special meaning for the purposes of this book. This means that ghosts are not *real*. In other words, they are not endowed with energy. This is one reason not to be afraid of ghosts since without energy, they cannot harm anybody. However, people indeed can see ghosts. That does not mean that the person standing next to the observer can see them too. Neither can they be photographed. (They have no energy, therefore, also no mass that could reflect light to form an image. Similarly, the ghost has to appear transparent to the observer.)

As we pointed out before, the "feeling of I" is an assembly of information bytes that form an image of the soul that is located in *existence.* The brain is the receptor for these information bytes coming from *existence* into *reality*. This is also true for the person who died hundreds of years ago. Although his brain is certainly destroyed and along with it the *feeling of I*, however, the information bytes that were sent to his brain to form the *feeling of I* still rest in *existence*. The interesting thing is that a part of this assembly of information bytes must have stayed together in the *Region of Existence* after death of the person. We may speculate here that we believe that this might happen, although we had no basis for such a guess. The fact that certain persons see ghosts would indicate that at least some information bytes that formed an "I" stay together. However, that may only be the case in extraordinary circumstances, like a violent death. Going through reports of what people "see" as ghosts, one could shed some light on this. The results of such an investigation could have a significant input on what we believe concerning life after death.

Why can certain people see a ghost? They certainly do not see it with their eyes. The eye requires light to see something, that is, energy, and ghost have no such energy. Certainly, the person sees the surrounding scenery. We have to assume now that this surrounding scenery contains a number of information viruses, which only this particular person who observes the ghost can read while the person standing next to him cannot and, therefore, does not "see" the ghost. By information virus, we mean that there are fragments of information in the scenery that the person who is now the ghost has left behind while he was still alive. This may be as simple as a few stones he may have moved. Or if he scratched a Chinese symbol on a stone, you will not recognize it since you cannot read it.

These information viruses together with what has been inputted into the brain of our ghost observer through ESP events form the image of the ghost in the observer. It is not seen with the eye, although the eye is needed to read the information viruses hidden in the scenery. This also explains why ghosts haunt certain places, namely, where the information viruses are located. Again, there never will be a proof for all these, and it has to be taken as pure speculation.

C. Spiritualism

Spiritualism is the activity devoted to communicating with dead persons. This is done by performing a séance, usually involving a medium. There are indeed a lot of believers; the cults practicing spiritualism are widespread. According to the claims made in the present book, a person's *feeling of I* consists of a large number of information bytes that are *realized* (meaning they reside in the brain and are as such subject to destruction after death). Albeit we forget some of the information bytes during our lifetime, the majority is destructed when the brain ceases to function. We have stated that all these bytes always existed, and always will, in the *Region of Existence*. If spiritualism ought to be practiced, it would mean to communicate with the *Region of Existence*. This would mean that at least some of the information bytes that have constituted the *feeling of I* of the dead person would have to be *rerealized* by the brain of the medium. Since information possesses no energy, any "signal output" has to be generated with energy provided by the medium. So far, so good. The issue now is do we believe that the information bytes that formed the *feeling of I* of dead persons stay together in the *Region of Existence* after death and can be accessed by a medium through some form of direct communication with the *Region of Existence*?

Again, if there is such a communication, it has to be like *intuitions*. Since, like in the case of intuitions, there is energy involved, and therefore in principle, such a communication channel could be detected by instruments. While such attempts may have been made, they certainly were not successful. To be successful, the researchers need to look for very fast bursts of extremely small voltages. The technology to do this is certainly available. However, funding for this type of research may be hard to obtain.

When ESP is practiced to find, for example, the body of a murdered person, then sometimes one gives an object that belonged to that person to the ESP practitioner. It is obviously hoped that the object contains or represents an information virus. In case of spiritualism, such a practice is usually not employed.

Spiritualists usually try to find out unknown things. Most of the time, the "cited" dead predicts future events to happen, e.g., warn of impending danger. Based on this, spiritualism could be verified to be a workable option, if such predictions actually come true. Again, a substantial amount of work needs to be done. Considering that some police departments use psychics, it would be a worthwhile effort.

D. Clairvoyance

This means to "see" the future. In order to be seen, the future logically must already exist. In the present book, we pointed out that information is indestructible. The information on what could happen based on the laws of nature resides already in the

Region of Existence. But there are, of course, several possibilities for what could happen in the future, and only one of those will be *realized.* This means only one of the many possibilities for future events will actually be endowed with energy and •will, therefore, influence our lives. In this book, we also claimed that life influences the future outcome of events. This means that the (*real*) future is not predetermined. That may mean that nobody can foretell the future.

Having said this, we have to admit that there is also another side of the coin. The "experimental" evidence for clairvoyance is overwhelming, starting with the ancient oracle in Delphi and continuing through the Middle Ages where Nostradamus is the most significant figure, through modern times, where psychics regularly predict the future, e.g., Edgar Casey, as published in the tabloids.

We may be forced to admit that there is something to it. In Europe, there are entire areas, the inhabitants of which are known to be endowed with a "second sight." These are the "Spoekenkiekers" in northern Germany, in an area that contains numerous moors. Likewise, there are certain English people, also living around moors, who seem to be able to predict the future. However, the most convincing "evidence" is that every single person living remembers an incident when shehe had the feeling that a certain thing will happen, and it indeed happened. Some people dreamed of things that actually have happened. Therefore, we are obliged to deal with all these. First, a prediction of something to happen within the immediate future (e.g., seconds) seems to be the most prevalent mode of clairvoyance. Deja vu also falls in this category. Since each one of us has experienced this, it will be difficult to explain it away with coincidence or fraud.

As a far-out speculation, one could surmise that one communicates (involuntarily) with another person (through *intuitions*), and this person either knows what has happened or is influenced by some exterior event to direct the flow of energy in the particular way that we predicted. In this case, only one other person will be involved, who has only a limited possibility to change the outcome of events. Therefore, only the immediate future can be predicted. Any longer time into the future would require communication with more people than just one since these other people could undo what the first one did. Therefore, the short-time clairvoyance is just another form of ESP. It is also not unreasonable to believe that this is actually true. This "sense" may have evolved for reasons of survival. If some enemy lurks behind the bush ready for ambush, one would need this forewarning to survive.

Here indeed would be a field for research. Such predictions of events, which were fulfilled, could they be caused by one or a few persons? One could collect a mass of data by interviews and find out if there is such a trend. Obviously, if there were such, it is more likely that the prediction a few seconds ahead is easier than a prediction through centuries. However, since *intuitions* are involved, distance in space and distance in time are meaningless. So even the predictions of Nostradamus could be plausible in principle. However, the complexity involved in "seeing" possible events through centuries than picking the right ones as the ones that actually have been *realized* is, to say the least, staggering. Yet we should remember that we found in our description of the *regions of the world* that *existence* has some structure, albeit a tenuous one. Therefore, it could be possible that the future events are clustered in major trends, and a gifted person like Nostradamus may be able to sense these. Chances are, he will pick some that indeed will happen but pick also others that will not happen. Since the predictions are presented in a disguised way, only the correct predictions will be recognized; the others will stay a mystery. This is a very old trick in future predicting; the oracle in Delphi already practiced it. In summary, acceptance of clairvoyance would mean acceptance of the concept of *intuitions*.

E. Astrology

This widespread semiscience is also concerned with the prediction of the future. We call it a semiscience since astronomical calculations are used to determine where the constellations of certain heavenly bodies were at the time of birth of a person. It is only the conclusions drawn from these calculations that are nonscientific, meaning not based on a chain of cause and effect. Again, we have to face the fact that a large fraction of the population believes in such predictions. One is inclined to say that if there were nothing to it, such a large number of people could not be fooled continuously. For this reason, we have to deal with astrology and determine what can be said about it, based on our model of the world. Again, what was said above about clairvoyance is true here as well. In contrast to clairvoyance, astrology does not really try to predict individual events but rather tries to describe trends. The trends are supposedly determined by the constellation of certain heavenly bodies at the birth of the person to whom these trends (and characteristics) apply. One could conclude from this that astrology suggests that these heavenly bodies actually through some mystical force influence our individual lives. Some proponents of astrology actually subscribe to this while others argue that these heavenly bodies are only indicators, like a clock, telling what the ongoing trend is, without actually causing this trend.

The fact that heavenly bodies can influence humans can actually not be denied. For example, the human female menstrual cycle (usually considered to be on the average twenty-eight to twenty-nine days) is about a lunar month. To be precise, a lunar month (full moon to full moon) is 29.53 days. A sidereal lunar month (from location at a certain star to the return of this location) is 27.29 days. Ironically, a human pregnancy lasts 9*29.53 days. Obviously, this cannot be a coincidence; the Moon must actually have an influence on this. However, this does not necessarily mean that there is a secret force that the Moon exerts. The full moon is a domineering heavenly body. Obviously, there is an influence not only on women but also on dogs and "lunatics." It is an accepted fact that humans and other life-forms can be influenced by hypnosis. It is safe to say that we do not understand yet what the mechanism is. But it is practiced, and it is known that a blinking light helps. Interrogators of all nations know that a bright light shown on the subject helps make him say things he does not want to say. Not too surprising that a bright light on the night sky may exert a deep influence. Also, it is not too surprising that there is no other heavenly body that does have a noticeable (psychological) influence on our life. (There is one known exception, namely, the worship of a certain African tribe concerning the orbital period of the invisible companion star of Sirius. We are at a loss to explain this and so is established science. Therefore, we do what established science does in a case like this; we do not talk about it.)

Astrology is actually successful in many cases; whether this is by coincidence or by reading signals displayed by constellations is arguable. If we admit that Nostradamus could have made a number of correct predictions, a number larger than to be expected from coincidence, then we have to admit that astrology can do the same thing. This would be based on the belief that possible future events are in the *Region of Existence* in a presorted fashion. Now, however, there is the added assumption that the sequence of events that actually will be *realized* is signaled by the constellation of heavenly bodies. Based on this, one has to say that astrology stretches plausibility a little farther than near-term or even Nostradamus-type clairvoyance. But we say this only since we cannot see why the future location of large heavenly bodies should be influenced by anything. One can compute where Jupiter will be in a million years from now. This is indeed predetermined. If the realization of a possible train of events ought to be linked to the location of Jupiter, this would mean that most of what is going to happen is already predetermined. This is indeed true for events too big for life to influence. But even if this link indeed would exist, predictions by astrology would be useless for human beings.

F. Stigmatization

The best-known manifestation of stigmatization is the case of Theresa Neumann, of Konnersreuth, Germany. She died in 1962. She regularly produced the wounds of Christ on Good Friday. This case was well observed and documented. There is no doubt that this actually happened. Stigmatization is actually not uncommon in Catholic communities. There are other reported cases that could produce the wounds of Christ even on command, cause those to start bleeding or stop bleeding, and could shed bloody tears. Symptoms like this are not confined to religious circumstances. There are also reported cases where patients could produce wounds they suffered in childhood just by talking about them. “Fakirs” are known to be able to drive skewers and knives into their flesh without feeling pain and without causing bleeding. A most interesting case was observed in France where a wife of a man to be executed by beheading developed a bleeding ring around her neck at the scheduled time of execution. Rather than explaining this with mysticism, one is inclined to propose that these are cases of extreme hysteria. It is sort of known to the medical community, but by no means admitted, that the will of a person can influence bodily functions.

Obviously, this can be carried quite far by persons who have a strong inclination for such a mechanism. In case of evolution, the will or desire may have been one of the evolutionary forces for at least the higher forms of life in combination with cooperative mutations by chance. Orthodox science ignores all these data since they cannot be explained within the existing theoretical framework. If the facts do not cooperate with the theory, that is too bad for the facts.

G. Miracle Healing

There were miracle healings going on in modern times in places where the Virgin Mary appeared, usually to children. Such places are located in France, Spain, and what was once Yugoslavia. The official Catholic Church is hesitant to take a position whether or not this is genuine. For the location in Spain, where the Virgin among other predictions declared that communism would disappear, the church allows but does not require the faithful to believe that the apparition has taken place. Since religion is a matter of belief and this book is not about religion, this topic shall not be discussed here any further. However, the healings that took place at these locations are too many to be explained away by fraud or self-deception (placebo effect). Our position is that the same effect as described above (stigmatization) is at work here. It is known that placebos can actually heal not unlike a pharmaceutical. Since there is a strong will required, it will not work for everybody. However, it could be a variation of what is commonly described as hysteria. Of course, this word has a bad connotation. The medical community believes that it is a disease, but it could be just a gift that every human being has to some degree.

Appendix 9

Time

A. Initial Remarks

Time is important for the subject of our book. Time is also a most mysterious entity for human beings. Humans may be the only creatures that know that death exists and is inevitable. For this reason, they plan ahead. They anticipate another day will come and another spring and another winter. They believe time passes and that they will die when their time has come. Therefore, time is their basic ordering scheme. Naturally, they want to know what comes after their time has expired.

Religion teaches the truth by authority; science searches for the truth by logical deduction. Any institution teaching or searching for the truth will have to address the question aimed at the "after." Science began when time was measured the first time by a human being. Humankind learned to predict when the Sun will rise and how long it would take for it to set again. They learned to predict the phases of the Moon and even solar eclipses. They divided the time in hours, minutes, and seconds. Obviously, they also asked, "Will time ever end?"

Or did time ever begin? And if so, what was "before" and what will be "after"?

Before humans discovered Earth being a sphere, they had the same questions for the "world." And in this case, they meant Earth. Does the "world" have a beginning and an end? If it has an end, how does this end look like? Is it a wall? Then what is beyond this wall? Or will it be an edge, and will ships fall over this edge? Where will they fall to? Yet there is a difference in the priority for knowledge. Whether or not Earth is a sphere and whether or not ships fall over the edge if it is flat may have been of ultimate importance for seafaring explorers but not for the peasants left behind.

Time, however, is of utmost importance for everybody because death will come at some time for everybody. It is a basic human need to know what comes after death. Humankind seeks this answer from whoever claims to know the truth. Is science in a position to teach anything at all to this subject? The answer should be affirmative; using the theory of relativity, physics is researching time and its structure. That should enable science at least to have a position whether time ever began or whether time ever will end. *If time ever ends, is there then no presence of anything, including a potential life after death?* One can expect that an institution, which investigates the structure of a phenomenon, is automatically required to investigate if this phenomenon has a beginning or an end.

Is *time* possibly an entity like a river in which everybody and everything is floating with the same speed? Alas, one might say maybe this river has little eddy currents and vortices. If so, it also must have some places where the water is stagnant, where it does not flow at all. In order to get from one current to another or even into the stagnant places, the velocity going through space is the determining factor. Of course, the question arises: Where is this river? What does it contain? Where does it originate? And where does it end. If there were such a river, how can it be detected?

Instead of answering this question right now, we pose another question: how do you measure time? The astonishing fact is that time cannot be measured. In appendix 1 (energy), we learned that something can be detected (measured) only if energy is exchanged with the entity to be measured. We called such an exchange an *energy-conversion event*. But this is not how time is actually measured. It is measured with a clock. For simplicity, let us stick to a pendulum clock. Here indeed, several *energy-conversion events* happen. Potential Energy is exchanged to Kinetic Energy and then exchanged again back to Potential Energy during a full swing of the pendulum (see also appendix 1 on energy). We measure time by counting the number of such *energy-conversion events* that happen during the span of time, which we would like to measure. In a regular measurement—when trying to measure everything else but time and empty space—a (small) portion of the entity to be measured is transferred to the measuring instrument. In this case, it is understood that the transferred portion is proportional to the total amount of this entity. The relationship of the size of the fraction of the total amount is found by calibration of the measuring instrument.

In the case of time measurements, we do not measure a fraction of the *time interval* of interest. We rather count the *number* of *energy-conversion events* regardless of the size of the time interval of interest.

The implication is that each pendulum swing always lasts the same length of *time*. We tend to believe that this is indeed the case, but there is really no way of ever finding out if this is so since that would require a method of measuring time. All methods to measure time, not just the pendulum swing, consist in counting numbers of repetitive *energy-conversion events* that have no relationship to the size of the time interval that we are trying to measure.

As pointed out above, we "measure" time by counting the number of *events* that happen. (Remember, for the purposes of this book, we call energy conversions "events.") Consequently, does this mean that if there are no such *events*, time does not pass? For space to be *real*, it has to be filled with *energy*. But does this necessarily mean that events are going on?

Strangely enough, the answer to this question has to be affirmative. The reason is that it is an undeniable fact (not a theory) that mass that has a temperature consequently will radiate infrared radiation (sometimes also called temperature radiation).

Why is electromagnetic radiation an energy-conversion event? Because it is a phenomenon where a magnetic field disintegrates and so causing an electric field to be established and vice versa. According to our definition, this is an *energy-conversion event*. One form of energy is converted into a different form or energy. This happens many trillion times a second. Therefore, in order for time *not* to pass, all mass in the universe would have to be at (absolute) zero temperature. But if such a situation would ever come to pass, masses would still attract each other and would form a gigantic sphere. The present black hole theory would predict that in such a case, a giant black hole would be formed into which all mass would fall and disappear. It may reappear in a different universe in form of a big bang. But this is wild speculation with tongue in cheek. We brought this up here to be able to make a comment concerning the special theory of relativity (STR). This comment states, according to this theory, everything is relative—location in space, mass, time. The only entity that is not relative is the velocity of light. This means the velocity of light is the same in all direction, regardless if the light emitter or the light absorber or both move. Since light is energy, the theory necessarily also states that energy is not relative. (Since we claim here that energy is a substance that cannot be created nor be destroyed, this seems to be reasonable.) Certainly, the form of energy that is observed may be relative and, therefore, different for different inertial systems, but the total amount is the same in either case. This is one of the basic theses that the STR teaches.

Did we come now closer to find out if time will last forever? Yes, sort of. We are now convinced that time will pass as long as *events* happen, and that is as long as there is radiation (light, infrared, microwave, etc.). This will happen as long as there is energy in the universe. Can the universe ever disappear? Logically, only if energy disappears first. In this case, where would it go? It is a basic law of physics that energy cannot be created nor destroyed.

B. Consequences

Time is an ordering scheme that orders sequences of events in an order consistent with the law of causality. It sees to it that the cause comes before the effect. In appendix 1 (energy), we showed that mass (and, therefore, Potential Energy) will distort space (see also appendix 4 on Einstein's pool table). Kinetic Energy will distort time in a similar way. The Lorentz transformations depicted in appendix 3 above describe this distortion.

Since time cannot be measured, is time *real*? The definition for *reality* states that it has to be endowed with energy. We explained in appendix 1 that *time* is the home of Kinetic Energy (not Kinetic Energy itself). So is it possible to measure a *real* entity "time?" The answer is no. Also, the *real* entity "space" cannot be measured either.

Distances in space are measured by putting bodies (e.g., yardsticks) in this distance and counting how many of those fit.

Distances in time we measure by putting events into these time distances that are to be measured and counting them. Therefore, time itself does not "flow"; it rather is Kinetic Energy that flows. Also, space does not "extend"; it is rather Potential Energy that extends and so distorts space. *Yet these are claims that, in principle, never can be verified or disproved.*

C. A Somewhat Less Complex Story

Here, we try to make things now somewhat easier to comprehend and still not violate the basic principles of the theory of relativity. We claim that there is indeed such a flowing river that represents time, namely, the flow of Kinetic Energy. Energy is considered to be quantized. This can only mean it consists of small particles that move from one location in space to another. That is the river of time. When we try to measure Kinetic Energy, we define it as:

$$E_{kin} = \frac{1}{2}mv^2$$

In plain English, this is an instruction how to measure Kinetic Energy. Determine the mass (m) of the moving body—one could say the number of energized atoms—then multiply it with the square of its velocity (v) and divide this by 2. Fine. The term "velocity" means first that the mass moves and second that it take time to move. (Velocity is the movement in space divided by time.)

Everything on the right side of the above equation (mass, length, and time) is relative according to the Lorenz transformations. But the left side of the equation is absolute. Energy, the substance cannot be produced or destroyed. This is a basic law of physics, and the theory of relativity specifically states that the laws of physics are valid regardless in what laboratory system (inertial system) we operate. Therefore, what we determine as the Kinetic Energy with the above equation is a specific value only valid in our laboratory system. It is not the absolute value of the energy, albeit there is such an absolute value.

What was said so far probably did not help all that much. Therefore, let us try an example. Again, imagine a railroad that has large windows on the cars so that a stationary observer can see what is happening inside the cars. One of the cars has a pool table, and the passengers play pool while the train is moving. Now observe one event as a passenger. Such an event will be when one ball hits another ball head-on. In this case, the moving ball will stop, and the (resting) ball being hit will take off. One could make the statement that the Kinetic Energy of the moving ball was transferred to the resting ball, and it has this Kinetic Energy now.

Now look at the stationary observer outside the train. He does not see the other ball come to a dead stop. The ball, originally at rest, was never at rest; it rather moved with the same velocity as the train. The ball originally moving (in respect to the train) has now slowed down to the same velocity the train has. That means that the passenger in the train would measure Kinetic Energies for each ball that are different from what the outside observer would measure. This is fine. As pointed out above, the individual form of energy may be relative, but the total sum is absolute. One could say one would have to add the train's velocity to the calculation, and everything would be fine. This is correct, but Earth also moves underneath the train due to its rotation and also due to its path around the Sun, and of course, the Sun moves too. All we can say is that it gets too complicated to take into account all these, and besides if nothing were moving, not even the train, the energy transfer event would be still the same—the moving ball would stop, and the resting ball would take off.

(For sake of accuracy, please understand that above is only true for straight-line movement of the bodies involved. The curved motion caused by Earth's rotation and its path around the Sun was neglected.)

In any case, we understand that Kinetic Energy moves through space while Potential Energy is stationary. In order for a body to move from A to B, it has to cover all points in between. It cannot disappear at A and reappear at B because energy can neither be destroyed nor created (conservation of energy). Kinetic Energy is the movement itself. If there is no movement, then there is no Kinetic Energy. Once a body is moving—has Kinetic Energy—this Kinetic Energy also has to keep moving. The law of conservation of energy does not allow energy to be destroyed; it, therefore, explains why there is inertia.

We can move in space by applying *Kinetic Energy* to our body. To move in time backward would mean to remove *events* that already have happened. We see in appendix 6 (entropy) that some events seem to be reversible and some seem to be irreversible. If there were indeed reversible events, such could be removed. However, in appendix 1, we claimed that the fourth law of thermodynamics states that in each event, all three forms of energy participate. *Heat* is the third form, and it is radiated away, not to be recovered ever again. Therefore, Heat increases entropy and makes an event irreversible. That is the reason why one can move in *time only forward*. This is done by creating events or letting events (like heartbeats) happen.

Does time have a beginning and an end? If time is related to the flow of energy, it is present as long as there is energy. One may believe that energy was created by divine intervention and may so be uncreated again. This is a matter of belief. There never will be a scientific proof that this ever happened or did not happen. All science can say is that energy cannot be created nor destroyed given the present state of the universe and the present laws of nature. If creation of the world happened (e.g., the big bang), then based on the present laws of nature, a miracle must have taken place. That is exactly what religion teaches and science cannot voice an opinion on. Consequently, if energy was created by a miracle, then time was too. *Energy can only be uncreated by a miracle, and so the time that goes with it; otherwise, there is no end of time.*

One might believe that *infinity* and *existence* were created also by divine intervention. We do not disagree or agree with this. The effect of such a belief is that the creator of this must be outside his creation. Our statement in previous chapters was: "if there is a god, he must be in infinity." We also stated previously that *existence* must be the knowledge of God (if one chooses to the belief that there is a god). Therefore, whether or not God is inside or outside of infinity is, of course, a matter of belief, and one cannot argue about the validity of a belief. We feel just more comfortable to state that God resides in *infinity* while *existence* is his knowledge. These two entities—*infinity and existence*—have no beginning and no end. Christianity does not subscribe to "Götterdämmerung."

D. Conclusion

The point that this appendix was trying to make is that the river of time is the flow of Kinetic Energy. *This is the opinion of AT; official science does not teach this.*

Therefore, if energy was generated by the big bang, there was no time before the big bang since there was no Kinetic Energy before the big bang. If there should be a catastrophic disappearance of energy (an anti-big bang), time will also disappear. Regardless of this, the *Region of Existence* will not disappear since it does not contain energy and, therefore, also not time. Therefore, if there are indeed souls, they must be in the *Region of Existence*, and consequently, *they would be indeed eternal.*

Glossary

Conservation of Energy, Also First Law of Thermodynamics

One of the basic laws of physics is that the amount of energy in the universe is constant. This means no new amount of energy is generated, and no amount of energy is destroyed. This is a grand statement. We really cannot prove that there is not a distant corner somewhere in the universe where energy is generated or that there is not a distant black hole in which energy falls and disappears. Maybe or maybe not. What is more important for us is that in the small vicinity where we live, no energy can be generated or destroyed. Even this would not be of extraordinary interest for everybody. What really counts is the case when one form of energy is converted into another form of energy. Remember, life thrives on energy conversions. At any conversion from one form of energy into another form of energy, the total amount stays constant. Sure, if we are trying to convert Potential Energy into Kinetic Energy, not all the Potential Energy shows up as Kinetic Energy. Some is converted into Heat, but the total amount of energy is equal to the original Potential Energy.

In order to make this more meaningful, take, as an example, your car engine. The chemical energy in the fuel is a Potential

Energy. When combustion takes place, this energy is converted into pressure, heat, and light. The pressure (still Potential Energy) is converted into Kinetic Energy when the piston is moved by the pressure. The heat and light hit the cylinder walls and is cooled away by the cooling system of the car. If all the energy forms are added up, it turns out that the sum of all is equal to the chemical (Potential) energy of the fuel used to operate the engine.

Energy

The book states that energy is a substance, even the only substance there is. All other substances, e.g., chemical compounds, are just various forms of energy as indicated by the equation $E = mc^2$. This equation states that energy, no matter in what form it is in, is endowed with mass. Logically, matter cannot be converted into energy since it is energy already. Therefore, the name of the equation is "mass-energy equivalence," not "mass-energy equality."

However, there seems to be some confusion. Some practitioners believe that it is indeed possible that mass can be made to disappear, and energy can so be generated. For example, the term "mass defect" was created for nuclear fission. It is believed that some mass of the uranium nucleus disappears, and about 200 MeV energy is obtained instead. That would be true only if the fission fragments, while having Kinetic Energy, had the same mass as if they were at rest. The "mass defect" is found by adding up the fission products (these are the fission fragments at rest), and indeed there is some mass missing compared to the mass of the original uranium nucleus. Therefore, one would have to state that the energy-mass equivalence is valid only for Potential Energy and not for Kinetic Energy. This is, of course, not true, and the "mass defect" is just a result of confusion.

The missing mass is now in form of Heat being converted into electricity in the nuclear power plant.

The book claims that there are three forms of *energy,* namely:

Kinetic Energy,
Potential Energy, and
Heat (Random Energy).

There are usually more forms of *energy* discussed. But all these forms can be fitted in above categories. For this book, it is important to understand that time is the host of Kinetic Energy and space is the host for Potential Energy. Heat connects the two together.

Energy cannot be created nor destroyed. This is known as the *conservation of energy* or also as the *first law of thermodynamics*. Energy can be converted from one form into another form. These are *energy conversions*. In technology, the most used conversion is from Heat into Kinetic Energy.

The fact that energy cannot be destroyed is most important for this book. *It points out that there are entities in this world that are eternal.*

Entropy

➢ Entropy seems to be related to energy, but whatever it is, it is not energy density. Originally, Sadi Carnot discovered that for his steam engine, a hotter steam works better. He knew already that he could not convert all the Heat Energy into Kinetic (mechanical) Energy but only a fraction of it. This fraction is still called the Carnot efficiency. It turns out that this efficiency is better the hotter the steam is. Now it was time to give this new phenomenon a name; it was called entropy. At any energy conversion, the entropy has to increase; otherwise, the conversion will not happen. Entropy was defined as $dS = dQ/T$

(Q is the original amount of heat). This seems to be a simple enough formula. The symbol T represents the temperature at which the amount of heat dQ was converted into Kinetic Energy. The *d* in front of the also *S* means a difference. It signals that there is no absolute value (e.g., zero) for entropy; only differences in the value of entropy are meaningful.

As pointed out, a process will only work by itself if the entropy increases. Therefore, the heat amount difference dQ has to be positive, meaning the Heat is extracted and converted to Kinetic Energy. Pretty soon, it was clear that entropy must be valid for any energy conversion, even if there is no temperature involved. Therefore, a new definition was needed that was valid for all energy-conversion events. The new definition is

$$S = k*\ln(P).$$

(*k*: Boltzmann's constant, *P*: the probability to find a system in this configuration; *ln* is the natural logarithm). From this definition, several facts can be gleaned. First, "to find a system in a certain configuration" means that all systems have a configuration. One can so say that *P* is the degree of order a system is in. *P* cannot be zero; neither can a logarithm be zero. This feature also establishes that entropy is not a substance like energy is. Something that cannot be zero cannot be a substance. Entropy rather measures the state of order a system is in. By "system," the thermodynamicists mean an assembly of objects, preferably in a box, that do not exchange energy with the environment (meaning anything outside the box). Such a system can be full of energy. The objects have some temperature and may even collide with each other. If left alone long enough, all energy inside the box will be gone. Where did it go? It was radiated away. Why? Because radiating the energy away was the only way to increase the entropy in the box. If radiation would be avoided by perfect insulation of the system, all objects in the box would eventually have the same temperature. Consequently, an entropy increase could no longer take place. If one surmises that energy conversions are necessary for life, one could say that the contents in the box would have died the entropy dead.

Equations and Numbers

Again, it is not possible to teach advanced mathematics in a few paragraphs. The following rather is intended to give an impression what the equations, used in the book, mean in plain English.

First, the matrix of numbers that explore a possible fourth dimension. Please refer to figure A6 on proportions of a possible fourth dimension.

The top row in figure A5 is 0D 1D 2D 3D 4D, which means zeroth dimension (yes, *zeroth* is no proper English word), first dimension, second dimension, third dimension, and fourth dimension.

The next numbers are all written as exponents. For example 2^4 is a different way of writing 2 x 2 x 2 x 2. Again the *x* here is the sign for multiplication. Therefore, 2^4 equals 16, and 2^3 equals 8. Similarly, 3 x 2^2 equals 12. In the figure, it shows that a solid cube has 3 x 2^2 = 12 one-dimensional entities (edges).

It should be pointed out here that the book does not claim that there is actually a fourth dimension. It could be that there is one, but if this should be the case, our semantic levels would not be able to comprehend it.

Next go to the following equation:

1. a = n * a
2. ∞ = n * ∞
3. 0 = n * 0

The "*" sign means multiplication.

These equations were made up by AT to demonstrate the meaning of infinity. The first equation is only correct for the case when $n = 1$. The second equation is correct for all values of n and so is the third equation. The common use of equations is that for any entity, only *one* Roman or Greek letter is used. If one sticks to the common use, a multiplication sign is not used. For example, *na* would mean n*a. Circumference of a circle is then written as $2\pi r$ (r means the radius, π means 3.141 . . .). The hypotenuse c of a triangle is related to the two sides (a and b) as described by the theorem of Pythagoras. This relation is described by the equation $c^2 = a^2 + b^2$ (provided that the tangle between a and b is 90 degrees). Again, c^2 means c times c.

Next the equation to describe a circle mathematically is

$$r^2 = x^2 + y^2.$$

In mathematical terms, r is the radius vector, and x and y are the ordinate and abscissa when a circle is plotted as a diagram.

Another equation used in the book is the instruction how to measure the Kinetic Energy of a moving body:

$$E_{kin} = \frac{1}{2} mv^2$$

Here, E_{kin} means Kinetic Energy. Since a multiplication sign is not used, the Kinetic Energy can have only one Roman letter, namely, *E*. If it is desired to be more specific, a subscript can be used, and these letters do not count. The letter *m* means mass, and the letter *v* means velocity. Therefore, the instruction for measuring the Kinetic energy of a moving body is measure the velocity *v* and multiply it by itself, then multiply by the mass *m* and divide it by 2.

The next equation is the most famous one:

$$E = mc^2.$$

E means any form of energy; *c* is the velocity of light (3.0×10^8 meter per second) and multiply it with itself and then with the mass (in kilogram). If these measures are used, the result is obtained in joule. Since *c* is such a large number, one obtains a substantial amount of energy even for a relatively small mass.

The next equations used are the Lorentz transformations:

$$t = \frac{t_0}{\sqrt{1 - \left(\frac{v}{c}\right)^2}}$$

$$m = \frac{m_0}{\sqrt{1 - \left(\frac{v}{c}\right)^2}}$$

$$x = x_0\sqrt{1 - \left(\frac{v}{c}\right)^2}$$

The symbols mean: m: mass; *t*: time; *v*: velocity; *c*: velocity of light; *x*: the distance in direction of movement. The subscripts of zero mean the value before movement. The large mathematical symbol is a square root. The "root" of a square root is the number that if being multiplied by itself has the value that is inside the square root. For example, the square root of 4 is 2 because 2 x 2 is 4. There is a slight complication because minus 2 multiplied by itself gives also 4 (not minus 4).

In order to understand above equations, there is no need to worry about the square roots. What is of interest are only the extremes, namely, if they are zero or infinity. These equations describe what happens according to the theory of relativity if a body moves in the *x* direction. This body needs to be inside an inertial system. Such an inertial system could be a spaceship that moves freely in space. It is an inertial system if no outside forces are applied to it. An outside force would accelerate the spaceship. The above equations apply only for continuous (nonaccelerated movement). Inside the spaceship, nothing would change if it moves. The value of *t* or *m* or *v* would not change. However, a measurement from the outside would reveal that the time passes slower inside the spaceship than the time of the measurer passes.

This can be seen from equation for *t*:

$$t = \frac{t_0}{\sqrt{1-\left(\frac{v}{c}\right)^2}}$$

For example, if *v/c* were 1/10, then *c/v* squared would be 1/100; therefore, what is left inside the square root is somewhat smaller than one. If v/c were one, then the square root would be zero. A division by zero is undetermined. However, one could use a very small number, not very different from zero, for example, 1/1,000. The result would be t and would be very large. In other words, an event that lasts one second at "rest" could take a thousand second if the spaceship is moving with a velocity close to the speed of light. Therefore, the time in the moving spaceship passes slower. To draw the conclusion that the time will not pass at all if the ship moves with the speed of light is, therefore, not unreasonable. Why can the ship not move with the speed of light?

The next equation shows this:

$$m = \frac{m_0}{\sqrt{1-\left(\frac{v}{c}\right)^2}}$$

If the ship were to move with the speed of light, the square root would be again practically zero, and the mass of the ship would become very large, practically infinitely large.

The third equation produces something different:

$$x = x_0\sqrt{1-\left(\frac{v}{c}\right)^2}$$

If the ship were to move with the speed of light, $(v/c)^2$ would be one, and the initial length of the ship x_0 is multiplied by a square root that is zero.

There are other equations in the book that need to be mentioned. They are not related to the theory of relativity. One is the DeBroglie wavelength :

$$\lambda = \frac{h}{mv}$$

Here, h is Planck's constant, which is a very small number. The symbols m and v are the same as above. The wavelength, λ which is the wavelength an electron or any other particle, would have to have if it were a wave (Schrödinger wave). Such a wave is required for explanation of the outcome of certain experiments.

Another equation used in the book is the following:

$$S = k \ln(P).$$

S is here used as the symbol for entropy; k is used for Boltzmann's constant. P is the probability for finding a system in a certain degree of order, and *ln* is the natural logarithm. The order of a system decreases when the entropy increases; therefore, P has to be defined accordingly. This is somewhat inconvenient. At one time, one wanted to define a neg-entropy that would decrease when the order decreases. However, this idea did not make it.

Finally, we discuss the two equations for the uncertainty relation:

$$\Delta E \Delta t = h$$
$$\Delta p \Delta x = h.$$

The symbol Δ is used to designate a difference. However here, since *h* is a very small number, Δ has to be a small difference in *E* (energy), *t* (time), *p* (momentum), and *x* (linear distance). Note that the multiplication signs were, as customary, omitted. Showing the multiplication signs would result in: $\Delta E * \Delta t = h$. These equations claim that none of the named entities can have an absolute accurate value. They are all of some uncertainty. Originally, one would expect that this could be expressed in an equation like this: $\Delta E * t = h$. However, it is important to note that the uncertainty can be different for each of the two factors. For example, *E* could be fairly large, and *t* could be very small as long as the product of the two is about equal to *h*. In the book, we use this feature to explain the transfer of intuitions from the *Region of Existence* into the *Region of Reality*.

Higher Semantic Levels

One may believe that logical thinking can only be accomplished if there is a language. This may be wrong. It is possible that a dog is capable of some minor logic reasoning, but it certainly does not do it using a language. For the purposes of this book, we assume that a language is necessary for logical reasoning. We do this because it helps make the message of the book more easily understandable.

Now lets us define what higher semantic levels are. For example, how do you explain to a blind person how the color green looks like? One has to admit that this is impossible. In other words, although the blind person has a language, this language does not contain the entity "green." Herhis semantic level is not high enough. For normal persons, it is not possible to describe the entity "fourth dimension" (if there is one). Well, it is possible to describe mathematically how many corners a four-dimensional body should (see also appendix 2) have, but we cannot comprehend where these corners should be. Our semantic level is not high enough. We cannot even discern whether or not there is a fourth dimension.

Information

For the purposes of this book, the word *information* has a special meaning. It means that everything that can be described is information. The description can be in form of mathematical equations, words, or letters. These descriptions have to be of consistent logic. For example, one cannot describe the process of fitting a square object into a corner of a round room.

Analog to the *Region of Reality* that is filled with *energy*, the *Region of Existence* is filled with *information*. In contrast to energy, information has no mass and is not affiliated with time. Consequently, the ordering scheme of the *Region of Existence* is not time and space but logic consistency. We understand now why energy is a substance while information is knowledge. We also should realize that the border between the *Region of Existence* and the *Region of Reality* has to be fuzzy. A part of this border needs to coincide with the surface of a body that consists of mass. Out of a clean surface of a material body, electric fields stick out, and electrons come out and return to it. The electric field diminishes with $1/x^2$. This means they never will become zero. They may be zero for all practical purposes. If we remember the *uncertainty relation*, we will understand that it is possible that some energy for a certain short time intrudes into the *Region of Existence* and so can pick some information. The so-modified energy transfers this information into the *Region of Reality* (our brain) by a burst of energy so short that our consciousness cannot recognize that such a transfer has happened. Nevertheless, some information was transferred, and we recognize this as an intuition.

Large Numbers

This part of the glossary is aimed at readers who are not familiar with large numbers and unusual equations; it is not aimed to teach advanced mathematics. It is intended to give some indications what the number and equations used in the book mean.

When going into the atomic level, numbers used to describe everyday events are no longer sufficient to tell the story. For example, 1 mole (22.4 liters) of air contains $6*10^{23}$ molecules. This strange number 10^{23} is an alternate way to describe a large magnitude that is a 1 having 23 zeros attached to it. Correctly written, this new number is

$$6.00 \times 10^{23}.$$

Here, the *x* symbol is the multiplication sign.
If one wishes to divide this number by 2, it appears as

$$3.0 \times 10^{23}.$$

Also, if one wishes to divide 3.00×10^{23} by 4, the result will be

$$7.5 \times 10^{22}.$$

This alternate way to write large number is very helpful if arithmetical operations need to be performed, yet what it means is still incomprehensible to the human mind. A million dollars would appear as 1.00×10^6. A billion dollars would be 1.00×10^9 (in the USA), and a trillion dollars would be 1.00×10^{12}. Make sure to understand that 10^{12} is not one half of 10^{24}; it rather is 1.00×10^{12} times smaller than 1.00×10^{24}. There is still a long way to go before the magnitude 1.00×10^{23} can be comprehended by the human mind.

Regions of Reality, Existence, and Infinity

We claim in this book that the world has three different regions, namely, the *Region of Reality*, the *Region of Existence*, and the *Region of Infinity*.

The *Region of Reality* is a three-dimensional entity that is structured by two ordering schemes, namely, *space* and *time*. Both of these are familiar to humans, but they are not really definable. Orthodox science may take exception to the claim that *reality* is structured by these ordering schemes. The Theory of relativity claims that the fabric of space is stressed by gravitational fields (obviously also electrical, magnetic, and all the other ones they may have dreamed up). The stress of the three-dimensional entity must affect, obviously, the structure of space. This is described by the Lorentz transformations (see also appendix 4 on Einstein's pool table).

The *Region of Existence* is structured less rigidly. This structure contains anything that can be *described but is not endowed with energy*. Such a description can be of any nature, an equation, a description by text, etc. The requirement is that this description is logically consistent (one cannot put a square object in the corner of a round room). The *Region of Infinity* has no structure at all, meaning it encompasses the other regions. Infinity is incomprehensible. It can be mathematically explored, but one has to realize that mathematics is applied logic (logic of humans). Mathematics cannot describe anything to humans that is not of human origin or not on a semantic level that is accessible to humans.

Time and Space

In this book, it is claimed that *time* and *space* are *ordering schemes* governing the *Region of Reality.* Although it is actually obvious that this is the case, you will not find this claim in most physics books.

We claim here that there are (at least) three forms of *energy*, namely, *Kinetic Energy*, *Potential Energy*, and *Heat*. Kinetic Energy has to do with movement. Movement in turn has to do with velocity. Velocity is defined as the distance traveled divided by the time required to travel this distance. The consequence is that *time* enforces the law of causality. It sees to it that the cause has to come first before an effect can take place.

Space also enforces the law of causality. It sees to it that cause and effect has to share at least one point in space at the same time for the effect to take place. In less esoteric terms, the ball on a pool table has to move (time) before it can hit the other ball. It also has to hit (space) the other ball for an effect to take place.

In most scientific measurements when a certain energy amount needs to be measured, a small fraction of this energy is detected (sidetracked), and this "measurement energy" is converted by an energy-conversion event into another form of energy whereby the energy conversion coefficient is known. Again in less esoteric means, if it is desired to measure how much electrical energy (measured in kilowatt-hours) a household drew from the grid of the electric company, a small amount of the current flowing into the house is sidetracked into a small electric motor (the meter). This small motor is calibrated, and the number of revolutions it performed is counted. In other words, a small fraction of the energy to be measured is sidetracked and analyzed.

In case of time, there is no way to sidetrack a small amount of the time to be measured. One rather uses events (for example, pendulum swings) and counts the number of swings that took place during the time of interest. The important point here is that the time required for a pendulum swing has no relationship to the amount of time to be measured. It is assumed that these swings take the same time each and every time. The point we are trying to make here is easier to understand if we assume that when in the electrical case the household draws too much current and consequently the little motor (the meter) is fed with too much current, it burns up. In case of time, there is no difference for the pendulum if it used to measure ten seconds or a century. It also does not matter what kind of pendulum is used as long as the calibration is done with a standard pendulum.

Similar arguments can be made for space. Distances in space can be measured with a ruler, but the ruler does not participate. It will not be ruined if the distance measured is too large.

Why do we go through the trouble to point this out in a book on life after death? The answer is that space and time are part of *reality*. If souls exist and have an image in reality in form of the *feeling of I* and if we claim that neither of these two has mass or energy, then we seem to have a problem. This problem is that we claim anything located in reality is endowed with energy and mass. And indeed the *feeling of I* is endowed with some energy, namely, the electrical potential of the cells that contain the code for the *feeling of I*. However, the soul itself cannot be in space or attached to a certain time. Consequently, the soul is not inside a person's body and will not leave this body at the time of death. The *feeling of I*, which is the bridge to the *Region of Existence*, is destroyed at the time of death.

Transfer Mechanisms

Transfer mechanisms are the ways and means of how to transfer information that is located in *existence* into *reality*. These means are intuition, information virus, teaching, reading, and others. In principle, intuitions would need to be considered paranormal. However, the present book claims that some scientific thought can be applied, namely, the uncertainty relation, to make intuitions less paranormal and more likely to be *real*.